UNIVERSAL INNOVATION

A Holistic Perspective on Innovation at Different Levels

Okan Geray, Ph. D.

Copyright © 2013 Okan Geray

All rights reserved.

ISBN: 1494359871
ISBN-13: 978-1494359874

Disclaimer: *Because of the dynamic nature of the Internet, any Web addresses or links contained in this book may have changed since publication and may no longer be valid. The views expressed in this work are solely those of the author.*

DEDICATION

To My Late Father,

Who incessantly challenged and

Inspired me intellectually and scientifically,

And,

To My Wife,

Who continually inspires me and

Plays an exemplary role in striking a balance

Between rationalism and mysticism

Table of Contents

Prologue 6
Innovation at Macro Scale – The Big Bang 8
    Expansion in the Universe and the Big Bang 8
    Fundamentals 11
    Early Universe – After Big Bang 18
    Initial Innovation after the Big Bang 22
    Innovation after the Radiation Era – Macro Structure Formation in the Universe 22
    Innovation through the Stars and Planets 25
    Some Intriguing and Enigmatic Facts about the Universe: Are We Living in a Rare Universe? 27
    Is there anything else in the Universe other than the visible matter? 30
Innovation at Planet Scale – The Life Bang 32
    Are we living on a Rare Planet? 32
    Evolution of the Earth in Brief 35
    Characteristics of Life 42
    The Life Bang after Big Bang 43
    Life and its Diversity over Time 44
    Cells as Building Blocks of Life and the Code of Life 47
    Signaling and Transduction in Organisms 58
    Adaptation of Organisms 59
    Connectivity Aspects of Biological Networks 62
    The Shape (Morphology) of an Organism 63
    Innovation in Life Forms (Organisms) 63
    Some Observations about Life on Earth 64

The Life Bang after Big Bang 66
Further Innovation at Planet Scale – The Technology Bang 68
    Observations about Technology 72
    Modeling Technology Tools 74
    Parts and Components in Technology Tools 77
    Modeling Innovation in Technology Tools 80
    Transduction and Hybrid Technology Tools 82
    Cumulative Innovation in Technology Tools 85
    Connectivity Aspects of Technology Tools 86
    The Shape (Morphology) of a Technology Tool 88
    Technology Tools - Purposeful Innovation 88
    Innovation Similarities and Differences between Life and Technology Bangs 89
Some Futuristic Organism Innovations – Foresight through Insight 93
    Demarcation between Organisms and Technology Tools 102
A Short Treatise on Other Innovations 106
Concurrent Innovation at Different Levels 112
Bibliography and Suggestions for Further Reading 122

Prologue

In most of my life, I endeavored with a childish curiosity to understand how all this visible material diversity that we see (and even we do not see which comprise the majority) had come into being and what the mechanisms were. Undoubtedly, my entire journey is not over but I have decided to write some of my own syntheses in this book. I have tried to put together a rational scientific perspective of selected theories, observations and thoughts in order to reach a personal unification. I have not discovered any of the scientific theories and findings mentioned in the book. But what I discovered was that when glued together, those findings and theories form an amazing collage of reality. In fact, they shed an interesting light on reality from a scientific viewpoint. In my humble opinion the scientific interpretation invokes deep inspirations for ultimate questions of reality and existence.

We live in a world of incessant change. Not a single day goes by without hearing the word "innovation". It is also frequently stated that the pace of innovation is accelerating in an unprecedented manner. This may be true when we refer to technology related innovations and their resulting products and services which we consume regularly in our lives.

However, I have taken a broader perspective on innovation in this book and explored its presence in different contexts and at different levels. This broader perspective of innovation informs us that large scale visible matter formations in the form of galaxies, stellar systems and planets take place over large timeframes such as millions and billions of years in the universe. Similarly, innovations of Earth itself and its various life forms on it have also taken place over the course of millions and billions of years. All these aforementioned innovations are highly dynamic and continually renew themselves, pointing to the fact that the only constant is the change itself in the universe. Physical laws play a fundamental role in setting the context and the landscape in which these innovations take place in the universe.

On the other hand, when we take a much smaller perspective by concentrating on the emergence of technology and its related innovations on Earth, we realize that they take place in much smaller timeframes such as years or thousands of years.

It is interesting to observe that innovations at the celestial, planet and technology levels possess certain common features and aspects. Innovations can be described as combinations of constituent parts in a hierarchical manner. Furthermore, innovations can be identified as either relentlessly fulfilling the fundamental laws of nature or also by fulfilling intended purposes of their designers. I will refer to the former as natural purposeless innovation and the latter as purposeful innovation. Technology innovations belong to the latter group.

One of the factors which prompted me to write this book is the recent advances occurring in genetics and synthetic biology. We, the human beings, are conquering the structure, mechanisms and internal workings of organisms in a hitherto unseen manner. For the first time in human history, we have attained a level of advancement whereby we can design novel organisms in addition to the naturally existing ones. There is a plethora of opportunities and challenges arising from this progress. Speculatively, it may define the next set of technology innovations on Earth in addition to the existing familiar ones such as information technology. We might potentially see a convergence of technology tools and organisms to meet intended purposes in the future. This is really an uncharted territory and we are merely experimenting and discovering it at the moment.

In a nutshell, this book can be considered as my personal discovery journey of innovation spanning various fields. It is a natural amalgamation of science, technology and engineering disciplines from a novel matter formation perspective which I referred to as innovation. It is also a quest for finding the simple principles behind innovation.

I have been fortunate enough to be present in exciting and innovative places throughout most of my career. Such environments and contexts certainly bless one with the opportunities to ponder about the fundamental principles and the underlying aspects of innovation despite the fact that in today's highly connected world location seems to lose its significance.

Okan Geray, Ph.D. - Dubai, 2013

Innovation at Macro Scale – The Big Bang

"Science fact is more interesting than science fiction"

Lawrence Krauss

The Nobel Physics laureate Jean Perrin once stated that the key to any scientific advance is to be able "to explain the complex visible by some simple invisible". The complex visible material forms that we see around us, both living and non-living, and without seeing how they formed might be blinding us to the simple invisible behind them. "Innovation through combination subject to natural laws" will be the common unifying theme of this book. This unifying theme will be recurring throughout the book in several contexts with scattered discussions. Formulas and scientific jargon are intentionally excluded from this popular science book in order to appeal to a larger audience (at least in order not to lose the interest and enthusiasm of readers who are not particularly well versed in mathematics and science). So let's start our journey from macro scale; that is, innovation at the universe level.

Expansion in the Universe and the Big Bang

Science is like a coin with two sides. It conducts observations on one side and formulates theories on the other side to explain the observed evidence. It is successful as long as the two are consistent and support each other. Successful theories can even predict the observations which sometimes are verified and validated in due course.

Scientific observations conducted by astronomers through telescopes indicate that we live in a gigantic universe with a very large number of objects in them. Earth is the most familiar object and the closest to earth is the moon which is 363,104 kilometers (225,622 miles away) at its closest point. In other words, driving at 100 km per hour we would reach the moon in nearly 151 days. Similarly the Sun is 146 million kilometers (91 million miles away) at its closest point. In other words, driving at 100 km per hour we would reach the Sun in approximately 61 thousand days. In fact the Sun is so far away that when we see the Sun we see its image 8 minutes ago because it takes 8 minutes for its image to arrive on Earth even though it is travelling at the speed of light. When we look at the stars at night, some of them actually do not even exist at the same moment we see them; because we see their old images belonging to an ancient time after travelling at the speed of light for millions or billions of years to reach the Earth. And in some cases by that time they may cease to exist. So essentially, the speed of light puts a barrier for us to observe the events. On Earth it takes much less than a second for the light to travel from one point to any farthest point so we observe events almost at the same time regardless of their distance on earth. But at large scales in the universe the same rule of thumb does not apply.

Advances in technology have enabled widespread ownership of digital cameras today. Taking pictures and videos of daily experiences and even sharing them instantly over the Internet have become commonplace. Despite the high penetration of digital cameras, a high ownership of telescopes to observe extra-terrestrial events and experiences is not so common. In fact, observations with telescopes have unraveled some of the most amazing mysteries of the universe (telescopes are amazing devices which allow us to look back in time; they create a collage of images from different times and present it now). We experience a similar time collage when we look at the night sky since each star is from a different time.

The entire universe is expanding at an accelerated pace so every celestial object in the Universe, including the Earth, is becoming farther away from all the other celestial objects in the Universe. This is referred to as the inflation of the universe or increase in space (analogous to inflation in economics which refers to increases in prices).

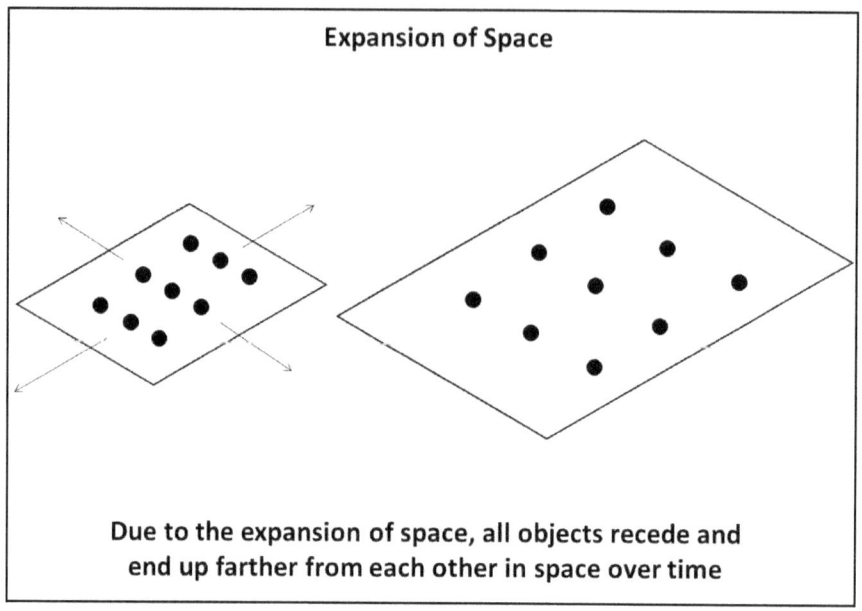

Figure 1 – Receding of Objects due to Expansion of Space

The expansion of the universe has interesting conclusions if we fast forward the universe at this expansion rate. In the future, close neighbors of Earth such as other planets and the Sun may not be as close as they are today (after all isolation might be a universal law that does not apply only to individuals). But it gives even more interesting results when we play it backwards in time. In other words, expansion means at any point in time the universe is larger than at any previous point in time. That is, if we move backwards in time, the universe becomes smaller. If we go sufficiently backwards in time and use the actual scientific observations and evidence, a very interesting conclusion is reached that the universe was tiny in its infancy about 13.7 billion years ago. Needless to say, if all the material and the energy we observe today are squeezed (collapsed) to a tiny point-like existence, then we end up with an extremely hot and tiny universe. And if we play it forward in time from then onwards, we end up with today's expanding universe.

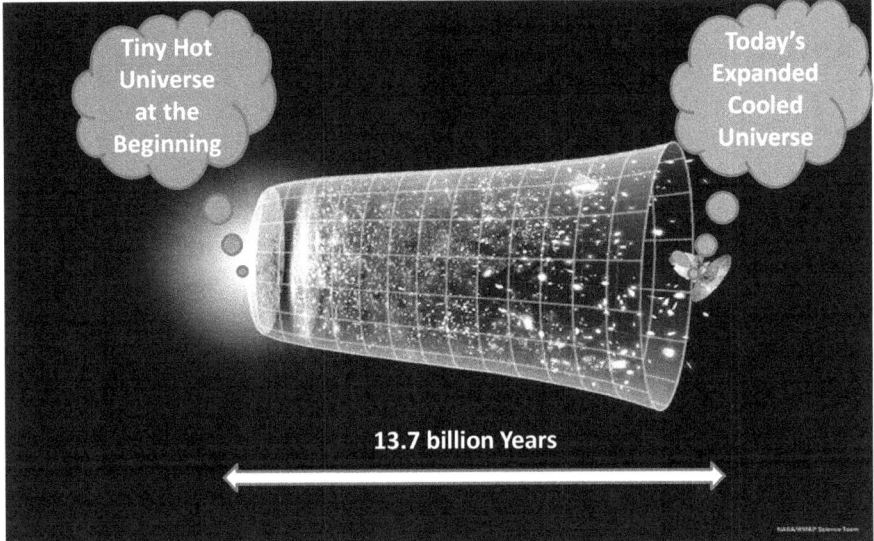

Figure 2 – Evolution of the Universe over Time (Image Courtesy of NASA/WMAP SCIENCE TEAM)

When this inflation (expansion) is plotted from the beginning till now with respect to time, it resembles an explosion and the after effects of it are as shown in Figure 2. This explosion is called the Big Bang theories in physics and the after effects are basically the evolution of the universe which is observed to date.

Before embarking on our journey with Big Bang, let's look at some fundamental aspects of the universe that we live in.

Fundamentals

The matter that we see in the universe has a very hierarchical structure and is composed of atoms (combination of atoms). Atoms on the other hand are a combination of nucleus and electrons, whereby electrons spin around the nucleus of an atom. Nucleus is a combination of protons and neutrons. Furthermore protons and neutrons are combinations of different types of quarks. Protons and neutrons both contain three quarks. So we see a very hierarchical structure for the atom (like an onion composed of different rings that can be peeled).

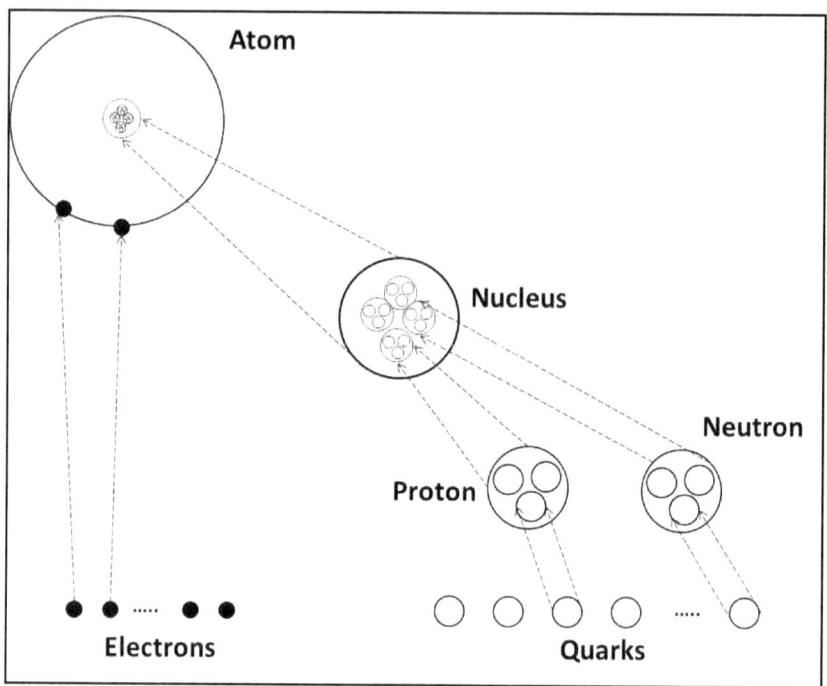

Figure 3 – Hierarchy of an Atom

Atoms having different number of electrons as well as protons and neutrons in their nuclei form different chemical elements in the universe. The ordinary matter that we see in the universe are either chemical elements or combinations of chemical elements called compounds (hierarchical as well). An atom is the smallest building block for a chemical element and molecule is the smallest building block for a chemical compound.

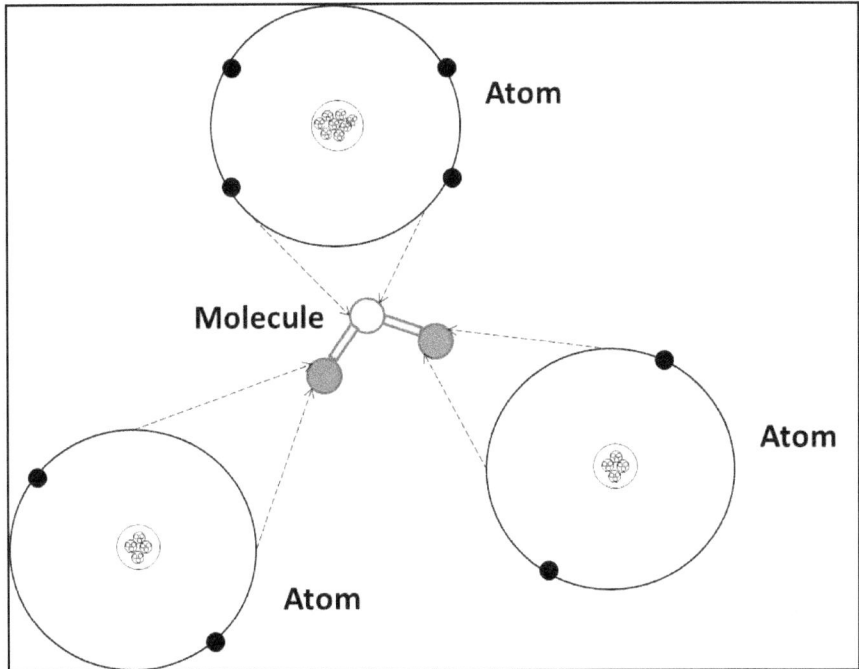

Figure 4 – Hierarchy of a Molecule

Figure 4 shows a sample molecule which is the smallest building block for chemical compounds. The representative molecule is made up of three atoms. As can be seen from the figure, it contains two similar atoms of a chemical element and one atom of a second chemical element.

If the smallest representative unit of matter, i.e. molecule, contains one atom then it is a chemical element, if the molecule contains more than one atom then it is a compound made of one or more chemical elements. The properties of a chemical compound are different from the properties of its constituent atoms; therefore a chemical compound can be considered an innovation, i.e. a novel matter type exhibiting different properties from its constituents.

In general, we can consider the ordinary matter around us as an innovation through combination in a hierarchical manner simply because molecules are combinations of atoms, atoms are combinations of electrons and nuclei, nuclei are combinations of protons and neutrons, and protons and neutrons are combinations of quarks[1] [2].

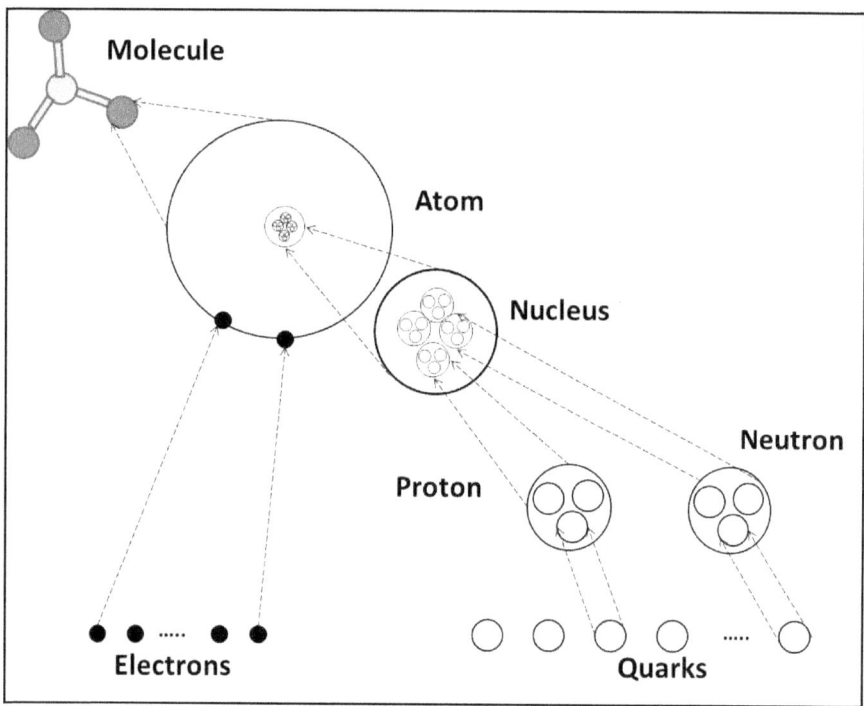

Figure 5 – Hierarchy of the Matter

The matter and its constituents interact with each other and these interactions are determined by the four fundamental forces of the nature; namely the electromagnetic force, the weak force, the strong force, and the gravitational force.

These forces have been responsible for the way the matter has formed hierarchically through various combinations of constituent parts both at the micro and also at the macro scales. Attractive forces play a combining role

---

[1] Some scientists strongly believe that this hierarchy continues even further. Our limitations to probe deeper in matter due to energy constraints hinder us from potentially identifying smaller and lower level constituent particles.
[2] In reality, there is a long list of family of particles in the universe with exotic names like fermions, baryons, leptons, etc. with each containing several particles in them. But to simplify things, we have concentrated on the ordinary visible matter we observe around us and analyzed its constituents. We have also omitted dark matter, antimatter and dark energy from our discussions.

whereas repulsive forces play a detaching role for constituent parts of the matter and for the matter itself as well. At the end, the four fundamental forces hold everything together and are indispensable for understanding the matter in the universe.

The Electromagnetic Force: Electromagnetic force determines the interactions between charged particles. In general, opposite electric charges attract each other and the like charges repel each other. It is the electromagnetic force that attracts or repels these electric charges in nature (universe). Electrons, quarks and protons are subject to this force but neutrons are not since they carry no charge. Inside atoms, electromagnetic force holds the negatively charged electrons in orbit around the positively charged nuclei. It is also pervasive between the atoms of a molecule and even among the molecules themselves.

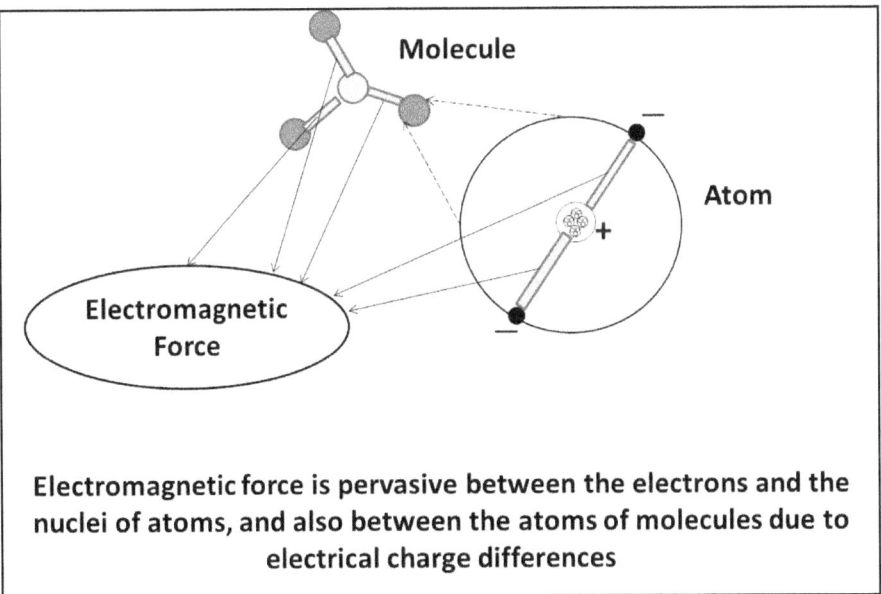

Figure 6 – Electromagnetic Force

The Strong Force: The nucleus of an atom consists of protons and neutrons. This is slightly enigmatic because we know that protons have positive electric charge and should repel each other in the nucleus. The protons in the atom's nucleus do indeed interact through electromagnetic force, but another force also operates between the protons and neutrons.

This force is much stronger than the electromagnetic force, so the latter's influence becomes negligible inside the nucleus. This is an attractive force, and it keeps the nucleus together. This force is called the strong force. The strong force operates between quark particles and do not affect the electrons since they are not composed of quarks.

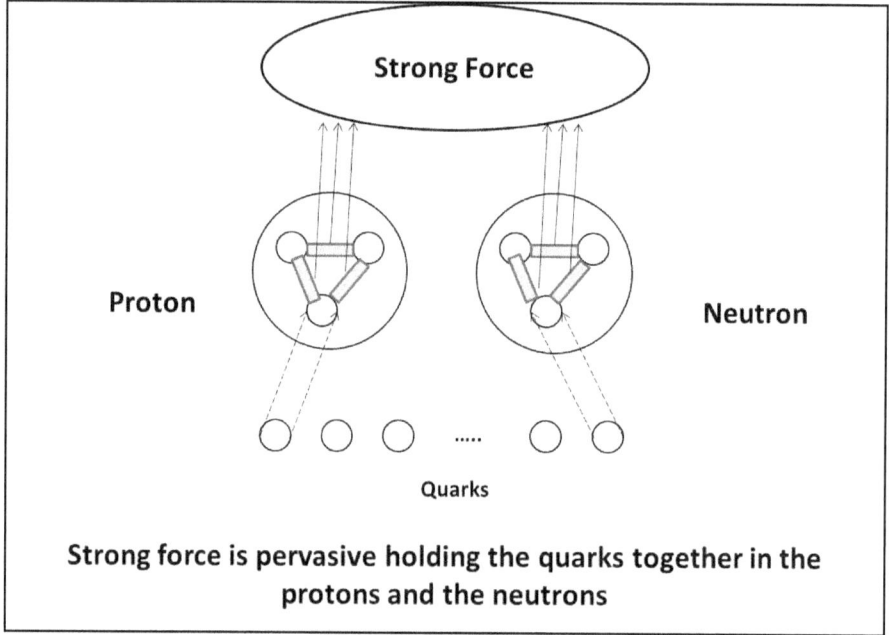

Figure 7 – Strong Force

The Gravitational Force: All masses (objects) in the universe attract each other and it is the gravitational force that causes this attraction. The gravitational force is responsible for dispersed matter to coalesce. It plays a fundamental role in planet, star and galaxy formations in the universe. It is responsible for keeping the Earth and the other planets in their orbits around the Sun; for keeping the Moon in its orbit around the Earth; and for the formation of tides.

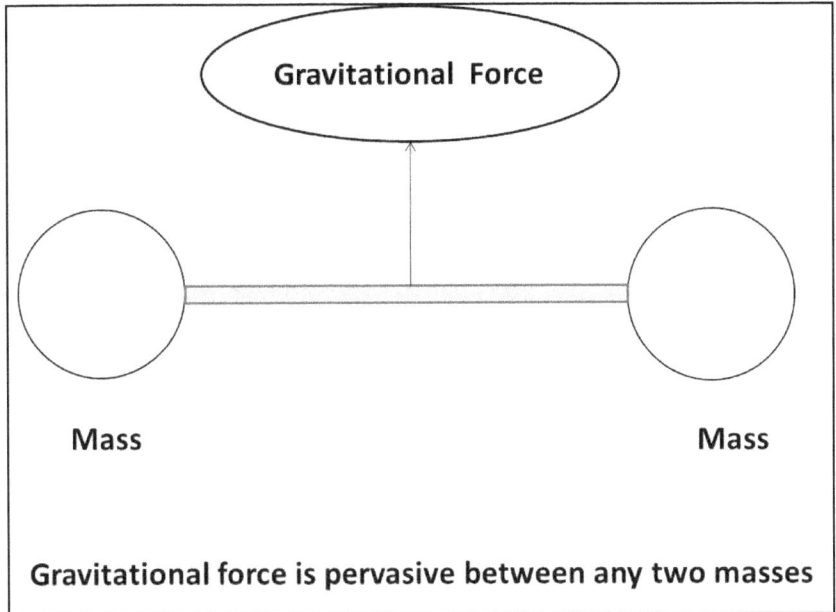

Figure 8 – Gravitational Force

The Weak Force: Weak force influences all quarks and electrons and is responsible for some kinds of radioactive decay. As its name suggests, this force is rather weak. The weak force has a very short range. It can operate only over distances that are approximately one hundred times smaller than the size of a proton. Two particles must be at least that close if they are to interact via the weak force. If they are separated by greater distances, the particles will not notice the effects of the weak force.

Out of these four forces, strong force has relatively the highest magnitude (strongest force among the four) and is approximately one hundred times stronger than the closest in magnitude which is the electromagnetic force. However strong force has a very short range and it operates within a proton or neutron. On the other hand, electromagnetic force has an infinite range and operates at any distance. The weak force is one hundred billion times weaker than the electromagnetic force and operates at extremely short distances, even one thousand times shorter than the strong force range. Finally, gravity is by far the weakest force than all the other three and is even 10 trillion times trillion weaker than the weak force but has an infinite range and operates at any distance.

Even though electromagnetism is far stronger than gravitation, electromagnetic attraction is not relevant for large celestial bodies, such as planets, stars, and galaxies, simply because such bodies contain equal numbers of protons and electrons and so have a net electric charge of zero (they all cancel out). Nothing "cancels" gravity, since it is only attractive, unlike electric forces which can be attractive or repulsive. Therefore, only gravitation matters on the large scale structure of the universe. At large cosmic distances, the extremely weak strength of the gravitational force has actually gigantic impact. On the other hand, at short distances in the range of atomic scale, electrical charges do not cancel out and adjacent atoms and molecules interact with each other through the electromagnetic force between them. This is the root cause of chemical reactions.

The matter and the fundamental forces of nature that we experience now did not exist during the Big Bang. They all came into existence as events unfolded after the Big Bang; so let's have a brief look at them.

Early Universe – After Big Bang

Right after the Big Bang (the birth of our universe) in the very early moments, the universe was extremely hot and dense since all the matter and energy that we see in the universe today was squeezed into a tiny space. In fact matter could not exist at such high temperatures. In other words, the temperatures were so high that the forces that hold the matter together (enabling the hierarchical combinations discussed above) could not withstand the pressure exerted by high temperatures on them in the very early universe (temperatures reaching billions of degrees and more). So it was a very hostile environment for matter to exist (it is almost impossible for us to envisage this state of the universe because we never come close to such energy levels in our lives as human beings and also it is impossible to sustain our lives at such high energies). In fact, the physics of such extreme conditions are unknown and yet to be developed. At such high energies (temperatures), it was a radiation dominated universe in which matter could not exist in a stable form. The entire universe at the very early times was an interesting mixture of radiation and matter (in fact matter and radiation were fluctuating into and out of existence with each other; i.e. photons, the building blocks of radiation, were disappearing into particle and antiparticle pairs which in turn disappear into photons, etc.). The particles naturally moved much slower than the photons moving at the speed of light but the densely packed universe where collisions occurred between particles and photons very frequently was kept altogether in an ensemble which scientists call plasma[3].

Physicists also conjecture that the fundamental four forces of nature did not exist distinctly in the very early universe. They were all unified into a single force and started becoming distinct later on in the very early universe. This is due to the fact that the four individual fundamental forces that we have seen lose their identity at very high energies (very high temperatures) which was the case right after the Big Bang. So the forces start behaving in a similar fashion and become indistinguishable at very high energies. However in the energy ranges and temperatures that we experience on Earth, these four forces definitely act as distinct forces and we experience their effects in a distinguishable fashion in nature. That is why we refer to them as the four fundamental forces of nature despite the fact that their distinct nature vanishes at very high energies.

The extremely hot and dense universe started expanding at a stupendous pace and the self-gravity of the mass-energy in the universe acted as a kind of brake to this expansion only slightly slowing it down. Needless to say, the expansion concomitantly caused the temperature of the universe to start going down. As the temperature went down, the universe reached a state where by particles could start forming matter due to the four fundamental forces mentioned earlier.

Nucleons (i.e. protons and neutrons) started forming from quarks when the temperature went down to below 1 trillion degrees. Then nuclei started forming by binding protons and neutrons together. The force that holds them together is the strong nuclear force. Nucleus is the bound state for protons and neutrons. Hence the strong nuclear force attracted and bound the protons and neutrons together and the temperature had cooled down sufficiently to avoid ripping apart these bound nuclei. In other words photon bombardments to bound nuclei could not destroy them.

The circumstances and the number of certain ingredients for making nuclei

---

[3] Several intricate details are omitted in this section for simplification purposes. The reactions and various reactants involved in the early universe are fairly complex and require a reasonably good mathematics and physics background to grasp the fundamentals.

were such that the number of nuclei formed during the initial phase of the universe, referred to as primordial nuclei, were constant and stable since the temperature was decreasing and there was not enough energy to rip them apart. Hence, in the first few minutes after the Big Bang the light nuclei had all formed, namely hydrogen and helium.

As the temperature went down further, another important threshold was passed whereby electrons would bind to protons in the nuclei through electromagnetic force (positively charged protons and negatively charged electrons attracting each other combined and the temperature was not enough to rip the electrons apart from the protons). And the charge-neutral atoms were formed for deuterium, helium and lithium. Also very importantly, at that point there were no free charged particles left for photons to interact with (photons normally interact most strongly with charged particles). Hence, photons were not impeded by collisions and they started roaming around the universe without obstructions (they still do till today from that time onwards). Consequently, the matter and radiation (photons) decoupled after about 370,000 years from the Big Bang and the temperature of the universe was about 3000 K degrees.

The decoupled photons roaming around the universe since then are called the Cosmic Microwave Background Radiation. The amazing fact is that we have an image of this very time when the radiation decoupled from matter as shown below (we could not obtain an image of the earlier universe because photons, or light, were locked in by the particles they collided with in the early dense universe).

Figure 9 – Cosmic Microwave Background Radiation *(Image Courtesy of NASA/WMAP SCIENCE TEAM)*

Scientists have estimated that due to the expansion of the universe this radiation would cool down after 13.7 billion years (i.e. now) to about 2.7 K degrees (the expansion of the universe causes the wavelength of Cosmic Microwave Background Radiation to increase over time thereby reducing the energy and hence the thermal blackbody radiation spectrum down to 2.7K). And this is exactly what they detected being present across the universe. A team of antenna designers in Bell Labs, New Jersey in mid-1960s detected a background wave (an irritating jittery noise according to them) which they could not eliminate from their antennae despite all the precision schemes they used. In fact, this was not noise but rather the Cosmic Microwave Background Radiation. They were receiving the Cosmic Microwave Background Radiation from the very early universe which fills out space uniformly. When they contacted scientists in Princeton University, they realized that without knowing they had discovered the Cosmic Microwave Background Radiation, in other words, the picture of the universe from the time when radiation decoupled matter. This picture is not just our earth's but the picture of our entire universe's origin. It was taken by an extremely sensitive array of telescopes. The color differences reflect the temperature variances in the early universe. In fact the color differences corresponding to the temperature differences are very minute (less than a millionth of a degree).

These minute fluctuations in temperature were imprinted on the deep sky (early universe) when the cosmos was about 370,000 years old. The imprint reflects ripples that arose during the very early universe. Astonishingly, these ripples gave rise to the present vast cosmic web of galaxy clusters and dark matter.

Initial Innovation after the Big Bang

The Big Bang was followed up by extreme conditions in the universe whereby matter and radiation coexisted in the form of an opaque fog, namely plasma. After 370,000 years, the laws of the universe created the initial *parts list* for universal innovation which consisted of only hydrogen, deuterium, helium and lithium.

Remarkably and intriguingly, we would have to wait millions of years to obtain the remaining chemical elements that we see around us. For example, water is one of the more common chemical compounds on Earth which is made up of hydrogen and oxygen atoms. The hydrogen in any water is in fact formed in the early part of the universe but the oxygen took millions of years to form.

The next section will talk about how the remaining elements were produced in the universe after the initial 370,000 years.

Innovation after the Radiation Era – Macro Structure Formation in the Universe

When the charge-neutral atoms formed, the radiation (photons) and the matter interacted with each other only gravitationally (i.e. under the influence of gravitational force among the four fundamental forces). In an expanding universe, the radiation density was becoming less than the matter density due to expansion when the temperature of the universe went down to approximately 2000 K degrees (after almost a million years from the Big Bang).

This is the matter era (as opposed to radiation era) where galaxies and galaxy clusters were formed including stars and planets. In this era, the matter under the influence of gravity started forming clouds of hydrogen and helium with some remnants of deuterium and lithium. Gravity normally pulls together (attracts) any form of energy including matter. If the matter is uniformly distributed in a precise manner in the universe then any attraction would be offset by the attraction with other neighboring energy. But any non-uniformity in the matter distribution will cause gravity to start lumping together nearby matter. The color differences that we see in the CMBR image are the evidences of non-uniformity in the energy (matter) distribution of the early universe. Therefore non-uniformity coupled with gravitational pull were the seeds of structure growth in the universe and led to the formation of galaxies and other structures due to gravitational attraction.

In the universe, we see a hierarchical structure whereby it is made up of galaxies and the galaxies are made up of stellar systems (stellar systems are composed of stars and planets).

Figure 10 – Galaxies (*Image Courtesy of NASA/JPL-Caltech*)

There are hundreds of billions of galaxies in the universe and the galaxies are also estimated to be composed of hundreds of billions of stars.

Figure 11 – Single Galaxy (*Image Courtesy of NASA/JPL-Caltech*)

Furthermore, there is interstellar and intergalactic material between the stars and the galaxies respectively.

Star Formation: The non-uniform matter in the universe was the instigator of star formation. It started combining (coalescing) and forming larger clumps under the attraction of gravitational force. Hence the elements and the particles (mostly hydrogen) that were produced after the Big Bang attracted each other through gravitation and started clumping towards each other, which in turn increased their temperature. After a few million years or so, this reinforcing cycle of temperature increase allows nuclear fusion reactions which are reactions whereby two nuclei join together to form a heavier nucleus. This is the basis of a young star formation. These nuclear reactions (also referred to as stellar reactions) result in the conversion of hydrogen to helium (massive stars can also form heavier elements) and release large amounts of energy from the star; i.e. literally burn out the star.

After this burning out period (which may last millions or billions of years), the star's nuclear reactions come to an end causing its core to collapse and the outer layers to be expelled as planetary gas nebula. Subsequently, depending on the mass of the star, it either explodes very violently (called a supernova) or releases its elements to the interstellar medium when its outer layer blows off. In the case of supernova explosion, even heavier elements are formed due to the presence of extremely high temperatures permitting stronger nuclear fusion reactions.

Planet Formation: Planets on the other hand are celestial objects orbiting a star (or its remnants) and massive enough to have its own gravity but not massive enough to undergo the nuclear fusion reactions which are characteristic of the stars. They can be either made of gas (called jovial planets) or solid materials (called terrestrial planets). The source material for planets can be the hydrogen and helium synthesized right after Big Bang and also heavier elements manufactured by the dead stars and their remnants.

Planets are usually born at the same time as their stars. As a star forms, a disk of gas and dust forms around it called a proto-planetary disk. In the direction perpendicular to the axis of rotation, centrifugal force resists gravity. Within this proto-planetary disk, planets will form. This disk surrounding a young star, with matter from the original knot continuing to fall inwards, is an accretion disk. It is also observed that young stars emit huge jets of matter. Therefore, matter falls towards the star (accretion) while simultaneously huge jets of matter are emitted. Therefore, planets form within the proto-planetary disk due to accretion and emission processes from the star.

Planets exhibit different properties depending on their chemical composition, and the distances from their stars among others.

Innovation through the Stars and Planets

It can easily be seen that stars transform themselves into energy and new elements by burning their fuels. Stars are in a never ending cycle of formation, burning and finally dying after they consume their fuels. In a way, they are the manufacturing plants (furnaces) in the universe for creating new and heavier elements which get spewed out to large distances in the interstellar space. Therefore stars enrich our parts list by producing other heavy elements in the periodic table.

In other words, stars have manufactured the heavy elements to be used in subsequent stars and in other structures such as planets. Without this constant recycling of stars, life as we know would not be possible since life is based on heavy elements such as oxygen, carbon, nitrogen, etc. which comprise the elements that make up our bodies. It is quite interesting to note that most of what we see on earth has been manufactured in stars somewhere in the universe and have been spewed out at unknown times by some unknown stars (at least unknown to us so far). There is literal truth in the adage "We are all star dust".

The stars have formed the foundation for further material diversity by manufacturing the heavier elements. However, ironically they themselves are very hostile environments since the temperatures in stars reach thousands or millions of degrees precluding the life as we know on Earth. On the other hand, terrestrial planets depending on certain circumstances can reach a certain stage after cooling which may be conducive to life as we know on Earth (the particular conditions and evolution of Earth as a planet will be discussed in the next chapter). Hence, stars are essential for material innovation in terms of enriching our parts list through heavier elements. But planets form the bedrocks (oases) for real innovation and material diversity by potentially consuming the parts list. Their relatively more affable conditions (compared to the stars) have enabled them as the preferred locations for the potential proliferation of life. Earth is an example of this verity and is teeming with life for billions of years.

When we look at the night sky, the celestial objects we observe seem static and stationary. On the contrary, the universe is immensely vibrant and dynamic by constantly renewing its celestial structures and objects over large scales of time. The galaxies and the stars collapse, merge, and acquire each other or they die burning their fuels resulting in catastrophic but astonishing celestial events at a large scale. However, these events occur in large time scales in the order of millions or billions of years so the universe seems deceptively and naively stationary to us.

Hence the universe is a riddle. Over the course of billions of years the unfolding of galaxies, stars and planets has enabled temporary structures in the universe. It is temporary because even to this day new stars and planets continue forming while the older ones cease to exist. It is a large-scale recycling of massive structures. They form, survive and cease to exist paving the way for new ones, which is their cycle of creation.

Some Intriguing and Enigmatic Facts about the Universe: Are We Living in a Rare Universe?

The parts list (atoms and elements) as well as large scale structures (celestial objects) are inherently dependent on the fundamental forces of nature.

- If there were no gravity, no stars and no planets would form since there would be no attractive force between matter to form them.
- If there were no electromagnetic force, no atoms would form (electromagnetic force enables protons and electrons to attract each other and co-exist in stable atoms or to form chemical bonds).
- If there were no strong nuclear force, protons would not co-exist in the nuclei of atoms since like charges repel each other (the nuclei of atoms would not form since protons would fly apart).

Like the fundamental forces of nature, there are also certain constants of nature (universe) which manifest themselves in various laws of nature.

- If the strong nuclear force constant were larger then hydrogen would be unstable and not form; conversely if the same constant were smaller then no elements other than hydrogen would form to a large extent.
- If the gravitational force constant were larger then the stars would be too hot and would burn up too quickly; conversely if the same constant were smaller then nuclear fusion would not occur and no heavy elements would be produced.
- If the electromagnetic force constant were larger or smaller then chemical bonding would not happen as it does in the current universe.
- There is a subtle balance between the expansion of universe after Big Bang that rips the matter apart and the gravitational force that pulls the matter together. If the expansion of space were larger then celestial structures such as stars would not be able

to form; conversely if it were smaller then celestial structures would form but quickly collapse.

The list above can be extended to include other constants and also ratios of constants in the universe rendering the current state of the universe highly delicate and sensitive to life formation. If any of the above happened then life as we know would not occur since above conditions rule out one or more pre-requisites to life formation.

High sensitivity to such parameters has divided scientists in terms of their explanations. Some claim that the sensitivity is just a coincidence in the universe. Some others claim that there is a high-level of fine-tuning in the universe, whereas some claim that we live in one of the many possible universes, referred to as the multiverse. The laws of nature are different in each universe in the multiverse and our universe just happens to be the one possessing these highly sensitive values which are conducive to life. This riddle has not been solved yet by the physicists. A counter argument can also be formulated such that there might be other laws of nature and different sets of constants of nature which may yield even better conditions for life to thrive in other universes. We seem to be lucky and find ourselves in a highly sensitive universe but also in one which has flourished life.

In this book, we will be more concerned about the outcomes of this universe rather than the processes that generate these outcomes. Hence, the laws of nature allow us to wind up with a sufficient list of parts produced in the Big Bang (light elements) and in the stars (heavy elements). The Big Bang and the stars are analogous to manufacturing plants that human beings on Earth use to produce different materials with the main difference that they require much more energy than our manufacturing plants and the manufacturing processes take much longer than ours.

Hence, the laws of nature leading to large scale cosmic structure (consisting of celestial objects) can be regarded as a natural purposeless innovation process. This natural purposeless innovation that only obeys the laws of nature has been continuing for billions of years and will do so in due course as well – our sun and earth will cease their existence just like other countless numbers of stars and planets did in the past and will do so in due course.

It is regarded as a natural purposeless innovation because the massive structures formed in the universe are not influenced or shaped by any individual being or any material entity (at least not discovered so far scientifically). Instead, the laws of nature determine the fate of the universe at a large scale.

Another interesting aspect of natural purposeless innovation is that it happens in large time scales and happens unconsciously[4].

The list of parts, i.e. chemical elements composed of atoms produced right after the Big Bang and later on in stars, can potentially be used on the planets to further facilitate innovation since they are the bedrocks (oases) for innovation and material diversity in the universe, with the Earth being a case in point.

Figure 12 - Depiction of the Hierarchical Structure of the Universe (*Images Courtesy of NASA/JPL-Caltech*)

---

[4] The above argument is from a scientific and human beings' perspective living on earth. Needless to say if the whole universe or portions of it are the result of a purposeful design then we are yet to discover its scientific evidence.

Is there anything else in the Universe other than the visible matter?

The hundreds of billions of galaxies with each containing hundreds of billions of stellar systems in the Universe is an indescribably gigantic amount of visible matter. Yet, scientists have discovered that this visible matter constitutes only 4% of the Universe. Nearly 26% of the universe is made up of dark matter and the remaining 70% by dark energy. The term dark essentially refers to our incapability for detecting them. In other words, we have no idea what 96% of the Universe is made up of. We have a good understanding of only the 4% of the Universe. Everything that we see around us, all the visible matter is just 4% of the entire Universe.

One important question that comes to mind is: So how do we know the existence of dark matter and dark energy? Dark matter cannot be seen directly with telescopes since it neither emits nor absorbs light or any other electromagnetic radiation. But its existence is deduced from its gravitational effects. The gravitational effects of the observed large scale structures in the Universe point to a scarcity in terms of the amount of visible matter present in them. Hence it is inferred that a different undetected type of matter (referred to as dark matter) should exist to compensate for this scarcity to make up for the observed gravitational effects. Dark matter is not made up of atoms as shown in Figure 3 for the visible (ordinary) matter. The exact nature and constituents of dark matter are yet to be discovered.

Similarly, the existence of dark energy is deduced from its negative pressure, i.e. negative gravitational effects. Dark energy is sometimes referred to as "cosmic acceleration". We have seen in this chapter that the Universe is expanding; however the expansion is also accelerating and its cause is unknown which is referred to as "the dark energy problem". Metaphorically, it is ripping apart all the matter in the Universe in an accelerated manner due to its negative pressure or repulsive effects on the matter. Just like dark matter, the exact nature of dark energy is unknown. The unrestrained expansion of the universe, if it continues incessantly, will separate celestial objects up to a point whereby they will not be able to detect each other in the distant future. The distance between celestial objects will be so enormous that light leaving one will not be able to reach another celestial object due to the huge velocity of expansion of space. It is quite ironic because we were ignorant of the existence of gigantic amount of visible matter in the Universe almost a century ago and we might be in a similar position in the far distant future again, but due to a different reason.

Consequently, we can state that we have a good understanding for the 4% of the Universe which is the visible matter. This book is about that 4% which takes almost 100% of our attention in our normal lives. The innovation that this book encompasses will deal with the visible matter which we see or detect. In the meantime, 96% of the Universe will remain as a mystery, waiting to be revealed.

Innovation at Planet Scale – The Life Bang

This chapter has a major home bias, which is our planet Earth. In the last chapter, it was indicated that the universe is composed of a vast number of galaxies, stars and planets. Unfortunately in this vast universe, Earth is the only planet for which we are reasonably well-informed. It is ironic because there are potentially more than trillion times billion planets in the universe however our access to them has been either limited (as in the case of the solar system) or virtually minimal, if not none. Therefore in this chapter Earth as a planet will be addressed including a brief in its evolution and the emergence of life on it.

Are we living on a Rare Planet?

This is a very difficult question to answer given our ignorance about other planets in the universe. At best our answer will potentially be biased towards Earth. However, there is also quite a bit of Earth related evidences and arguments which suggest that after all Earth may be a good candidate for a rare planet since it has enabled and supported complex life. So let's look at some of these.

- Complex life needs a variety of parts in our parts list introduced in the last chapter; i.e. it requires heavier elements such as carbon, nitrogen, oxygen, sulphur, etc. which are produced (manufactured) in stars. Hence, any planet that supports life should be in a region within the universe where such heavier elements are readily available (Earth already possesses these elements).
- Complex life on Earth needs water in the liquid state. Therefore a planet has to be within a certain range of orbital distance away from its star in order for the surface temperature of the planet to sustain liquid water. This range is referred to as habitable zone.

Figure 13 – Habitable Zone *(Diagram Courtesy of NASA)*

If the planet is closer to its star than the habitable zone, the water on its surface would evaporate and if farther than it would freeze Hence, a range of acceptable distance exists which can sustain liquid water on the planet's surface and that zone is referred to as habitable zone for a planet. As it turns out, Earth happens to be within the habitable zone for the sun.

On the other hand, we have seen that stars, including our sun, have their own life cycles and eventually die out after burning their fuels. Life on Earth will sooner or later be roasted out of existence. The sun is slowly getting brighter. It is now estimated to be 30% brighter than it was in the early history of the Earth. Over the next 4 billion years it is expected to double in brightness. Even if life on Earth endures this doubling of brightness, it will then face a more catastrophic event. About 4 billion years from now, the sun will begin to expand rapidly in size, and its brightness will dramatically increase. In approximately a further billion-year time span, its brightness will increase over 5000 times. At the very beginning of this process, Earth's oceans will vaporize, propelling Earth's water supply into space. In the final stages of its expansion, the sun will swell to the point where it will nearly reach the orbit of Earth.

Therefore, the next few billion years are the fortunate times to be on Earth.
- Complex life needs time to evolve. Life on Earth has taken billions of years to reach its current diverse and complex stage. Hence, the star of a planet has to also survive a similar long time period to support life on its planet. Life time of a star is related to its mass. The more massive the star, the shorter its lifespan, primarily because massive stars have greater pressure on their cores, causing them to burn their fuels (hydrogen) more rapidly. The most massive stars last an average of a few million years, while stars of minimum mass burn their fuel very slowly and last tens to hundreds of billions of years. Our sun's lifetime is predicted to be approximately 11 billion years which is sufficient to permit the evolution of complex life on Earth.
- A stable orbit for planets is necessary for life to subsist. It is well known that close proximity to giant celestial objects can significantly impact the stability of a planet's orbit. Nearby giant celestial objects whose orbits deviate from circles exacerbate the stability of a planet's orbit. Earth's orbit is quite stable thanks to the particular configurations and positions of nearby planets and other celestial objects.
- Outer giant celestial objects in a star system protect inner planets from various impacts. Earth as an inner planet is quite fortunate to have a giant planet like Jupiter which provides protection by gravitationally attracting a high majority of the asteroids and other outer space objects to itself, and thereby avoiding their potential collisions with the Earth. In a way, Jupiter acts as a heavenly shield for Earth.
- Earth tends to have the right size for sustaining its atmosphere. A small planet cannot hold on to its atmosphere since it cannot exert sufficient gravitational attraction. A planet without an atmosphere tends to have variable surface temperature and it also becomes more difficult to sustain its oceans. Earth happens to be in the right size and mass range to sustain an atmosphere, relatively stable surface temperatures and oceans.
- Earth has relatively a large Moon which provides a multi-fold stabilization effect on Earth. It happens to be at the right

distance from Earth and stabilizes its tilt. The right tilt provided by the Moon causes seasons not to be too severe and indirectly supports life on Earth.

However in distant future (couple of billion years from now) Moon will recede from Earth enough to alter its tilt (Moon is gradually moving away from Earth every year) and that is predicted to cause sudden and violent climate changes on Earth (this is another reason why we live in fortunate times on Earth now).

- Plate tectonics or the movement of the Earth crust across the surface of the planet is also crucial for life. In fact plate tectonics acts as our planet's global thermostat by recycling important chemicals through which the amount of carbon dioxide remains fairly consistent and that in turn enables water to remain liquid on the surface of the planet (has done so for more than four billion years). Plate tectonics also contributes significantly to Earth's magnetic field which protects the Earth from cosmic rays that would be detrimental to various life forms on Earth. Additionally it also contributes to maintaining diversity of life forms. Interestingly, Earth is the only planet in the solar system that possesses plate tectonics.
- Existence of carbon at the right amount has played an important role on Earth. Life on Earth is based on carbon containing compounds. Furthermore, carbon dioxide as a greenhouse gas has played a critical role in maintaining appropriate temperatures for life. Too much carbon dioxide might have caused boiling away of the oceans known as the runaway greenhouse effect (the same might have happened in Venus in its early history).

The above list can be extended to include even other chance events or characteristics which according to some scientists make Earth fairly exclusive for life as a planet.

Evolution of the Earth in Brief
We have seen in the last chapter how dynamic the universe was despite its

seemingly deceptive static appearance. Similarly when we look at our planet today, we see a fairly stable one consisting of continents, oceans and enclosed by an atmosphere. Yet this constancy is as deceiving as the stability of the universe since our observation timeframes are substantially shorter than the lifetime of the planet itself. The crust, oceans and the atmosphere have all gone through major changes throughout Earth's life and continue to do so as well.

Earth is about 4.5 billion years old, nearly one third of the age of the universe. As explained in the planet formation earlier, it formed from coalescing of particles (also called accretion) from solar nebula. Initially earth was mostly molten (a fireball almost liquefied by heat) due to extreme volcanism and recurrent collisions with other celestial bodies.

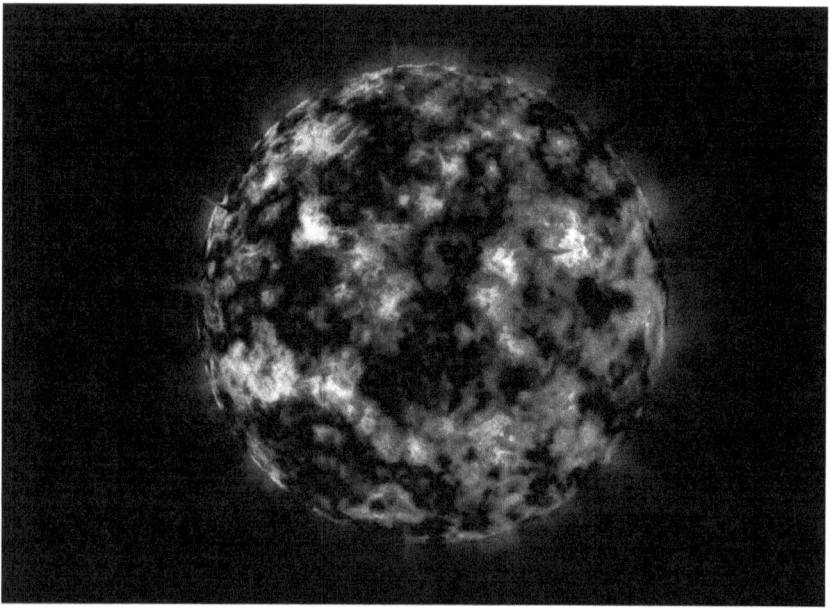

Figure 14: The Depiction of the Earth as a Fireball

In fact one of those collisions with almost a Mars sized celestial object is thought to tear away a good portion of the Earth which was ejected into orbit around the Earth. Under gravitational force, this ejected material became Moon later on and also tilted the axis of the Earth.

UNIVERSAL INNOVATION

Figure 15: The Depiction of the Collision of Earth with a Mars Sized Object

The collision between Proto-Earth and Theia created the Earth and the Moon 4,500-4,400 million years ago. Both planets had a massive iron core when they collided. It was a cataclysmic event at the time and tilted the axis of the Earth which had a benign effect in the long run by stabilizing the climate on Earth.

Changing Atmosphere: It is believed that the volcanism (outgassing) in early Earth also created its initial atmosphere which contained virtually no oxygen or ozone and would have been detrimental to most life forms on Earth today since it mostly contained carbon dioxide and nitrogen with trace amounts of other gases such as methane, ammonia, etc.

Today oxygen makes up approximately 20% of the atmosphere. So there has been a major shift from almost no oxygen to one fifth of oxygen in the atmosphere. This is referred to as the Great Oxygenation Event (GOE) which happened almost 2.4 billion years ago. In fact the main source of this oxygen was the bacteria (more specifically called cyanobacteria) which appeared 200 million years before that time. The photosynthesis of the cyanobacteria produced the oxygen (oxygen in carbon dioxide broken down into oxygen) and oxygen is well known to be a highly reactive element. The oxygen produced by cyanobacteria was captured by dissolved iron and also other organic matter on Earth. When there was no more iron or organic matter left to capture oxygen, the oxygen started accumulating in the atmosphere as a gas. Today this accumulated oxygen makes up 20% of the atmosphere.

Changing Continents: Not only the atmosphere but also the continents have gone through major changes during the history of the Earth due to plate tectonics. The single supercontinent Pangaea began to break up about 225-200 million years ago, eventually fragmenting into the continents as we know them today.

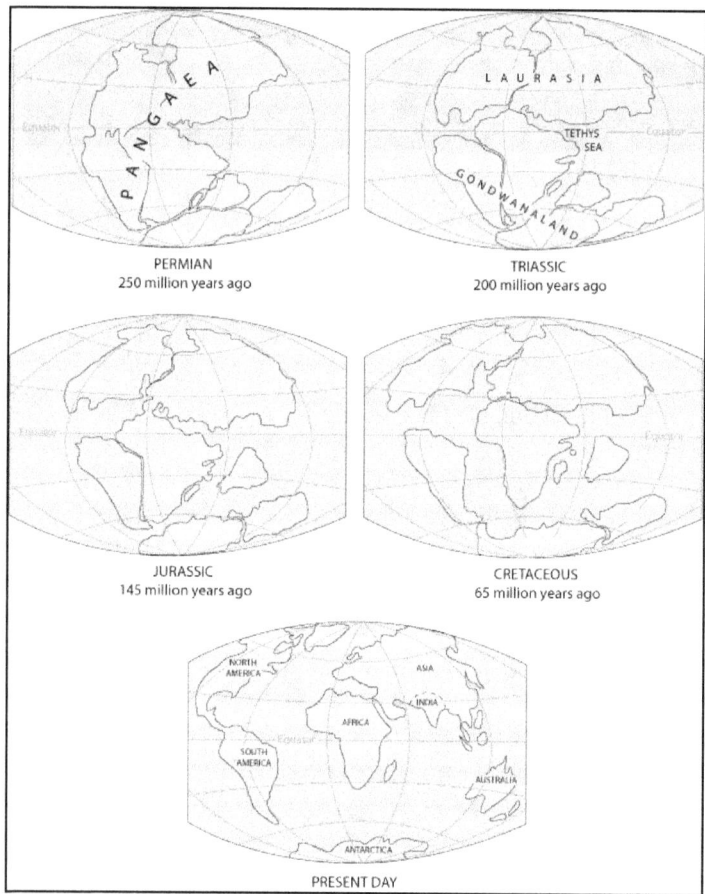

Figure 16 - The Break-up of the Supercontinent Pangaea (Courtesy of the U.S. Geological Survey)

Changing Oceans: The Earth initially was molten and did not have any ocean for millions of years. The volcanic activity on Earth also ejected large quantities of water from Earth's mantle and created the oceans. The active volcanoes in the early Earth ejected lava, ash and gases from deep within the Earth.

Figure 17 – The Depiction of Volcanic Activity in the Early Earth

One of these ejected gases was water vapor. Over millions of years, the ejected water vapor cooled and condensed to form clouds. These early clouds in Earth's history brought torrential rains which started covering the surface of the Earth and they formed the oceans in due course.

Figure 18 – The Depiction of Torrential Rains Covering the Surface of the Earth

Today, more than 70% of the Earth's surface is covered by water and the oceans contain 97% of the Earth's water.

Changing Surface Temperature: The Earth has witnessed several extended periods of temperature reduction in Earth's surface and atmosphere resulting in ice sheets and glaciers which are called ice ages. During the ice ages, there have been intermittent warm and cold periods.

Scientific evidence points to at least five major ice ages in Earth's history. The first ice age is predicted to be 2.1 to 2.4 billion years ago, the second occurred from 850 to 630 million years ago, the third from 460 to 420 million years ago and the last ice age started about 2.6 million years ago and it still continues. Arctic, Antarctic and Greenland ice sheets are a testimony to the current ongoing ice age. Of these five ice ages, the most severe one is predicted to be the second one whereby the glacial ice sheets covered the Earth entirely or nearly entirely including the equator. This is also referred to as snowball earth.

Earth Impact Events: These are events that were instigated by celestial objects causing substantial effects on Earth. Meteorites, asteroids and comets may potentially cause impact events after colliding with Earth. In fact such major impact events have influenced Earth in the past. We have already seen that Moon was formed after such an impact event caused by a Mars sized celestial object hitting the Earth.

A team of Canadian researchers from the National Institute of Astrophysics in Ottawa concluded that the annual amount of meteorite that falls on our planet is equivalent to 21 tons (there are different estimates from different sources). The size of meteorites falling on Earth range from that of a grain of wheat to the rocks that weigh more than a ton.

One of the best-known recorded impacts in modern times was the Tunguska event, which occurred in Siberia, Russia, in 1908. This incident involved an explosion that was probably caused by the airburst of an asteroid or comet 5 to 10 km (3.1 to 6.2 mi) above the Earth's surface, felling an estimated 80 million trees over 2,150 km2 (830 sq mi).

On 15 February 2013 while writing this book, an asteroid entered Earth's atmosphere over Russia with an estimated speed of 18.6 km/s (over 41,000 mph or 66 960 km/h)—almost 60 times the speed of sound at that altitude —the Chelyabinsk meteor—over the southern Ural region. Russian authorities stated that 1,491 people, including 311 children, sought medical attention in Chelyabinsk Oblast within the first few days.

In the past 500 million years there have been five generally-accepted, major mass extinctions that on the average extinguished half of all species. One of the largest mass extinction to have affected life on Earth was 250 million years ago and killed off 90% of all species; life on Earth took 30 million years to recover. The last such mass extinction led to the demise of the dinosaurs and coincided with a large meteorite impact which occurred 65 million years ago.

Figure 19 – The Depiction of the Large Meteorite Impact 65 Million Years Ago

The craters that we see on Earth are evidences of such impact events which happened in the past.

The changes we have seen on Earth (atmosphere, crust, temperature, etc.) were all natural events triggered by the laws of nature as they were applied to the conditions which Earth was under in those times. But they are testimony to the fact that Earth has evolved drastically over time and it was a very different place from what we see today. The changes we have seen did not only affect the Earth itself but also the life forms on it. In the next section, we will talk about the emergence of life on Earth and its evolution over time.

Characteristics of Life

The emergence of life on Earth was one of the most crucial events contributing to material diversity and innovation. Once again we will discuss specifically life on Earth since our knowledge of life on other planets is almost nonexistent.

Unfortunately there is no unique definition of life among scientists. In this book, we will use the important characteristics that describe life. Most definitions for life encompass a subset of the following characteristics:

- **Hierarchical Structure:** Life comprises living beings (organisms) which are composed of one or more cells (i.e. single / unicellular versus multi-cellular life forms). Cells are the building blocks of life. Unicellular life forms have a single cell. However, multicellular life forms such as plants and animals have combinations of different cell types which make up tissues and organs. Therefore structurally life is composed of cells and their combinations, which has a well-defined hierarchy. The living beings are distinguished from their environments and their cells are covered with a protective layer called membrane.
- **Metabolism:** Living beings (organisms) need energy and to transform that energy into other forms to sustain themselves. Life forms have devised different metabolism (energy consumption and transformation) methods depending on environment conditions and the availability of materials. Metabolism includes various chemical reactions that take place in living beings.
- **Growth:** Cells, the building blocks of life, grow in favorable environments (having nutrients) increasing their cell sizes and numbers.

- **Reproduction:** The ability of living beings (organisms) to produce new ones is reproduction. This allows continuity of life forms. There is one more important role of reproduction which is the transmission of information during reproduction (inheritance of information). The newly reproduced living being(s) carry the information needed for their own life processes (we will explain this further later on).
- **Adaptation to Environment:** Living beings have both short and long term adaptations to their environments. Short term adaptations are mostly responses to certain stimuli in the environment; examples might be cells producing certain chemicals in the presence or absence of other chemicals, movement of a bacterium towards a nutrient source or away from a toxic substance, plants turning their leaves to sun, etc. Long term adaptation is attained by collective organisms whereby inherited characteristics change over generations, i.e. genetic changes take place. Several traits and morphological (shape related) changes in species and organisms are long term changes due to adaptation of organisms to their environments. Unlike short term changes, long term ones have caused permanent changes in the genetic information existing in the cells of the organisms, thereby qualifying them as long term adaptations since they persist over time.

The Life Bang after Big Bang

Big Bang and large scale changes we have seen in the universe in the last chapter had created our rich parts list composed of light and heavier elements. Furthermore, we have seen that interactions among the members of our parts list, i.e. the elements, due to the fundamental laws of nature have created new chemical compounds and mixtures. That is to say, various combinations of the elements in our parts list due to chemical reactions had formed new materials. The matter that we see around us on Earth are mostly due to these chemical interactions that either happened naturally or that were produced through induced chemical reactions. However, those were all inanimate matter.

Origin of Life – A Riddle: But then in the history of Earth, a major chemical innovation occurred and life came into existence, which we call in this book as the Life Bang. We have seen that life has several characteristics which are quite different from inanimate matter. The origin of life and how it exactly started is a highly controversial topic. There are two different schools of thought regarding the origin of life on Earth; the extra-terrestrial one and the terrestrial one. Extra-terrestrial one alleges that the complex chemicals (possessing the characteristics we discussed) or even living cells came from outer space to Earth (carried by meteors, asteroids, etc.) and initiated life on Earth. The terrestrial one, on the other hand, alleges that the primordial chemical soup present on Earth have led somehow to the emergence of life after complex interactions. There is strong consensus among scientists in this category that particular or even certain precise conditions in highly specific environments might have contributed to the emergence of life on Earth. Such conditions and/or environments might be difficult to replicate (we do not know the exact chemical materials and the Earth environment when life arose). Needless to say, there are many different versions of both schools of thought which attempt to explain the origin of life with different assumptions and mechanisms.

As we have seen life has several characteristics. Different theories claim different hypotheses as to which characteristic appeared prior to the others. For example, some claim that metabolism appeared on Earth prior to reproduction, or vice versa. Some others claim that membranes appeared first prior to metabolism and reproduction. Therefore, it is widely assumed that there was a gradual cumulative build-up of the characteristics of life over time through chemical evolution, which finally led to the Life Bang, that is the emergence of Life on Earth.

Life and its Diversity over Time

Earth was formed approximately 4.6 billion years ago. After several hundred million years life emerged on Earth which in this book we called the Life Bang. Earth was a very different place in those times and evolved drastically in terms of its general environment (oceans, atmosphere, crust, etc.). The first life forms were unicellular (i.e. single celled) and remained so for more than 2 billion years. In other words for more than 2 billion years life on Earth existed as tiny little cells in the form of bacteria and alike. They are called prokaryotes by scientists. Their cells are simpler and do not have many specialized units or compartments including the nucleus as well. These early prokaryotes lived in the oceans where there was an abundance of chemical materials which they used as raw materials and sources of energy. The oceans protected them from lethal radiation (e.g. ultraviolet) since the atmosphere in those times did not contain oxygen and hence the ozone layer which later on acted as a shield for radiation. Hence, outside the oceans life could not exist due to the hostile environment of the Earth.

Prokaryotes have existed more than 3.5 billion years and hence devised different ways to consume and transform energy (metabolism). For example, we human beings (and various other animals) need oxygen (breathing) which triggers the metabolic reactions and we cannot survive without oxygen for more than a few minutes. On the other hand, prokaryotes have been able to adapt and devise different metabolism types using what was available in their environment as a source of energy (they were able to cope with changes in Earth's environment). These sources of energy include carbon dioxide, light, inorganic and organic compounds, nitrogen, methane, oxygen, etc. When it comes to energy consumption and transformation prokaryotes, despite their simplicity, are much more versatile than complex animals on Earth including human beings. Even though prokaryotes were very versatile in terms of their metabolism their morphology (shape) has not changed much throughout Earth's history. Some prokaryotes, known as archae, have been able to flourish in extreme environments on Earth such as deep thermal vents in oceans, frozen mountains, deep in Earth's crust, etc. Life has been able to thrive even in potentially hostile environments. Prokaryotes tend to grow rapidly and reproduce frequently.

Prokaryotes are like living fossils and have been around for a long time. They are the most plentiful life forms. In a simple glass of water, one can locate more prokaryotes (e.g. bacteria) than the human population on Earth.

About 2.7 billion years ago, photosynthesis (using carbon dioxide and light to produce oxygen and food) was devised by some prokaryotes as their metabolism. Hence, oxygen started forming as a result of photosynthesis and the unused oxygen started accumulating in the atmosphere. Oxygen was toxic for many prokaryotes rendering them extinct but those who could tolerate oxygen flourished and used it in their metabolism. Gradual accumulation of oxygen for about 2 billion years allowed a dense enough ozone layer to form approximately 500 million years ago which provided protection from ultraviolet radiation. Therefore life forms started migrating to land from the oceans.

In the meantime life forms innovated a novel cell type with membrane bound specialized units / compartments including the nucleus serving different functions. The new life forms with this novel cell type are called eukaryotes. Until over a billion years ago both the prokaryotes and the eukaryotes were unicellular.

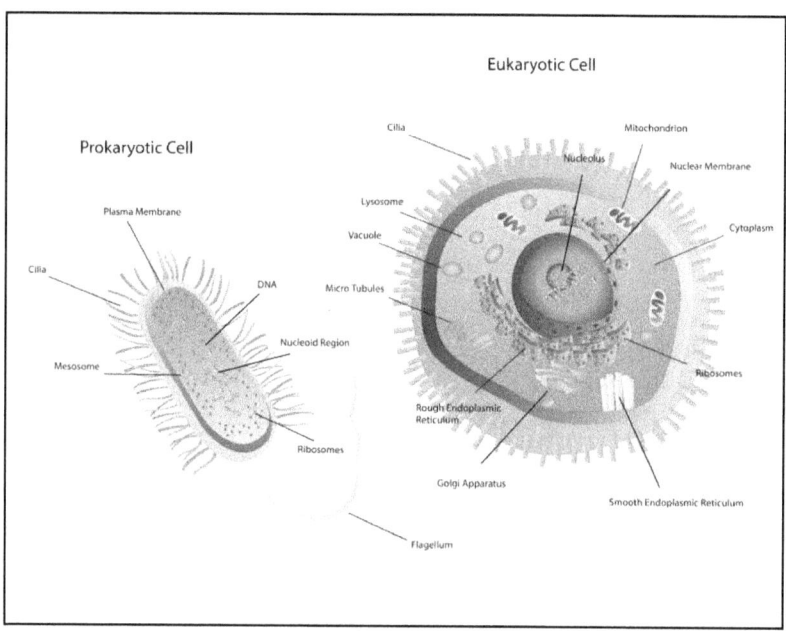

Figure 20 – Prokaryote and Eukaryote Cells

Later on, life innovated a novel living being (organism) type whereby some unicellular eukaryotes did not separate after cell division and the dividing cells remained together leading to multicellular life forms. This connected nature of cells enabled specialization of cells over time. In other words, different cells started performing different functions (e.g. some cells specialize in production, some in getting nutrients, some in processing them, some in transporting them, etc.). This specialization enabled multicellular eukaryotes to become larger in size and also to be more efficient in their functions.

Eukaryotes, unlike prokaryotes, have devised morphological (shape related) mechanisms (rather than metabolism related) to cope with changes in their environments. They have developed new body parts and diversified substantially. Initially plants and then animals of various sizes and shapes came into existence on Earth. 550 million years ago the diversification accelerated in animals which is referred to as the Cambrian explosion. New body shapes, parts and sizes came into existence fairly quickly (evident from fossils) relative to the previous history of life on Earth. This period is marked by the sudden appearance of larger fossils in size around the globe. Before Cambrian explosion, the land on earth was barren; the seas had worms and jellyfish but no sea creatures with wide variety or visible skeletons (no crabs, clams, shrimps, sea urchins, etc.). It was very different from what we see on lands and in seas today. Most living beings before the Cambrian explosion were very small and did not have skeletons but Cambrian explosion innovated novel larger life forms with skeletons as well. Around the Cambrian time, significant diversity of animals appeared within a time frame of 100 million years within the seas first. And then a similar explosion of diversity happens on land with plants and animals.

It is interesting to see that on Earth life forms did not necessarily evolve gradually, but rather long stagnant periods were followed by periods of abrupt changes in terms of diversity. From then till now, a staggering diversity of millions of animal species came into existence. The interesting fact is that more than 30 different template body plans have been identified in all these millions of different animal species. Depending on different combinations of certain parameters in these more than 30 template body plans, i.e. variations on a theme, numerous animal species with various body plans came into existence. Therefore, combinations of parameters in body plans resulted in body plan innovations in animals.

Cells as Building Blocks of Life and the Code of Life

The most fundamental unit of life is the cell. About a billion years after Earth's formation, nature has innovated life, i.e. the first cell by combining various materials in our parts list (toolbox). By definition, cells have the characteristics of life which we have seen earlier. Hence they consume and transform energy (i.e. metabolize) and they also store and transmit information (i.e. reproduce) and also adapt to their environment.

The first innovated cell incorporated very important information. To be precise, a very important code existed in the cell that provided instructions to the cell in order to perform its functions. The code was stored in the cell in a chemical form and was in discrete (digital) format; i.e. the chemicals in the code represented one of the four possible letters and the code consisted of a long set of letters (i.e. chemically bound as a chain together). We can think of the cell as a piece of hardware (the cell, i.e. the hardware, was organic as opposed to electronic) and the code as a set of instructions, i.e. software. Hence, the cell as a hardware building block comprised a platform whereby different instructions could be performed depending on the code residing in the cell. And life throughout billions of years used this simple mechanism to prosper and to flourish. Prokaryote and eukaryote cells were the two types of hardware building blocks (platforms) used by life. On the other hand, the software or the code consisting of instructions varied drastically over the course of billions of years. As expected, unicellular (single celled) prokaryotes and eukaryotes had a simpler code, i.e. set of instructions, since they were relatively simpler life forms. On the other hand, multicellular eukaryotes especially complex plants and animals had quite complicated codes, i.e. sets of instructions, which allowed them to conduct the life functions (e.g. metabolism) in the face of changing and challenging external environments. In a way, the life codes in them have become robust over time allowing them for example to change their body plans as we have seen in the previous section.

In other words, life was a chemical innovation which had created a piece of chemical hardware (i.e. the cell itself or a combination of cells, pieces of hardware, in the case of multicellular life forms) and a piece of chemical software for performing various functions. And over the course of Earth's history, life thrived on its chemical hardware while enhancing its chemical software gradually and sometimes quite abruptly as well (e.g. Cambrian explosion). The code or the software contained instructions which told the cell what chemicals to produce and when depending on certain external and internal conditions with respect to the cell. These produced chemicals, in turn, allowed the cell to carry out its functions.

There was one more important property of the cell as a building block of life. It replicated itself. This replication included both the hardware and the software (the code) so the code was never lost. Cells grow by division (replication) and each time a cell divides (replicates), it forms a copy of itself including its hardware and software. This allows the new cell to also produce chemicals that allow it to perform its functions. The machinery for replication was there in the first cell during the Life Bang (reproduction as one of the characteristics of life).

Sometimes, certain codes (or chemical software) did not produce benign results for the cells, for example the produced chemicals reduced the chances of the cell to survive under certain environmental conditions (in some extreme cases this caused the extinction of the corresponding life form) or made the cell to grow abnormally (we call this cancer in our daily lives). So the code was extremely important for the cell to function properly.

In multicellular eukaryotes, cells started specializing as we have seen earlier. In those cases, individual cells based on their proximity to other cells and the internal and external conditions with respect to the cell itself, would produce different chemicals even though the hardware and the software was the same for all the cells. In other words, the production of chemicals was regulated based on certain cellular conditions. The production of different chemicals would allow the cells to specialize despite the same hardware and software. This allowed individual cells to innovate in terms of their production capability. Specialization also required different cells to interact with each other (cell-cell interactions). So a cell would produce a chemical which would then be transported to other cells for usage or storage (e.g. hormones in human body, oxygen and food being transported through the blood, fat cells being stored in human belly, etc.). So the same code in every cell potentially produces different chemicals and at different times based on different conditions which allows specialization of cells. The produced chemicals are transported between the cells allowing communication (signaling) between them. For example, insulin is produced by the cells of the pancreas which removes excess sugar from the blood. Insulin causes cells in the liver, skeletal muscles, and fat tissue to absorb sugar from the blood. Examples like this are abundant in organisms.

The chemicals, which the code (software) in the cells produce, are called proteins. Every life form or living being does all its functions through the proteins produced by the code in the cell. Proteins are combinations of other chemical materials called amino acids. There are more than 20 different amino acids which can be combined in myriad ways. Amino acid combinations are called proteins. For example we can combine a few amino acids to make a protein or even hundreds of them. And each amino acid combined will be one of the more than 20 different types. So basically life processes are based on a combination of more than 20 different types of chemicals in a repetitive manner and in different numbers.

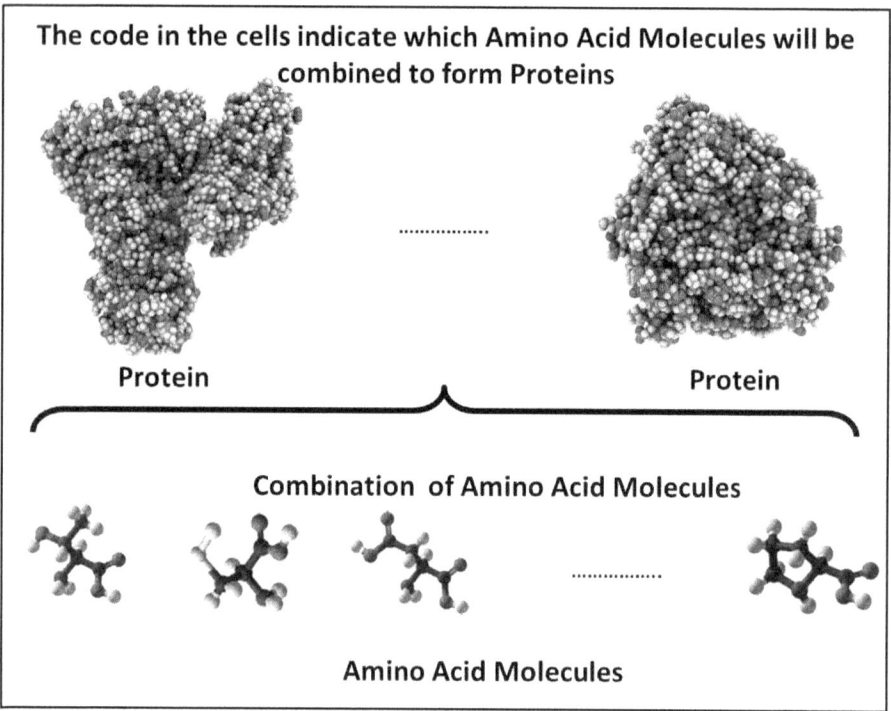

Figure 21 – Proteins as Combinations of Amino Acid Molecules

The code in every cell essentially indicates which amino acids to use in what order and how many of them. Literally, in the code within a cell, there are many sub-codes which have a START sequence and then indicates which amino acids and in what sequence to be combined to make a protein and then ends with a STOP sequence. It is just like a Morse code or a piece of software; and it is called a gene. So the code contains many genes depending on the life form (organism) and each gene indicates which protein to produce within the cell. Some produced proteins directly start a chemical reaction in the cell (e.g. metabolism reactions), some may hinder other reactions on purpose, some may accelerate or trigger other reactions, etc. So the proteins produced by the code play different roles in the cell. There are on the average approximately several hundred amino acids in a protein (needless to say the actual number varies from tens to hundreds to thousands by protein). Therefore, the genes are part of the code in the cell and provide the precise instructions to produce proteins by indicating which amino acids to combine in what order.

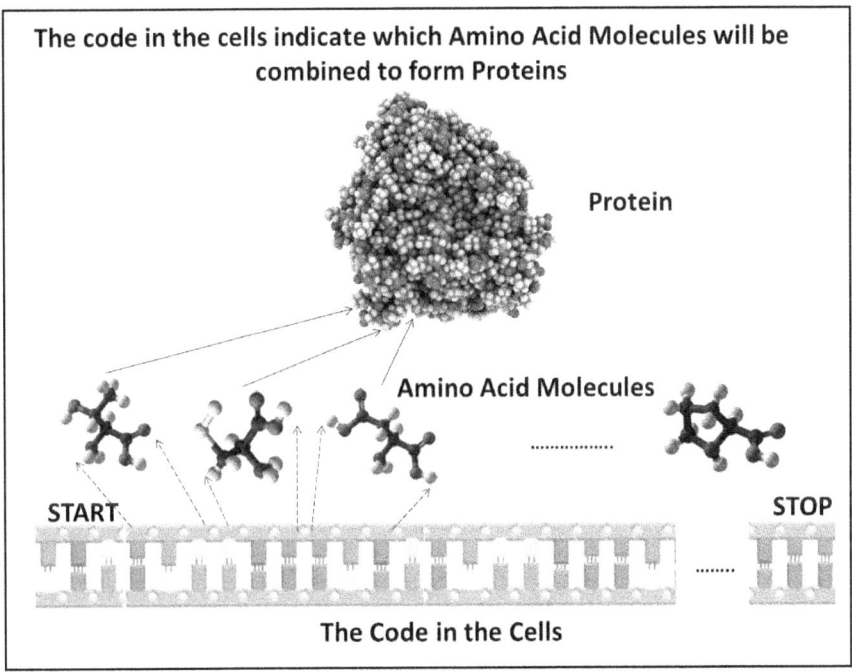

Figure 22 – The Code in the Cells and the Protein Assembly

The cell as a hardware unit of life comes with the machinery (the chemical manufacturing facility) to assemble and to produce the proteins indicated by the code (software). As indicated before, there are fundamentally two different cell types, prokaryote and eukaryote cells. The bacteria are prokaryote life forms and on the average include 3000 genes in their code (software) which produce proteins (i.e. their life processes are based on production of 3000 different proteins). On the other hand, animals may have 10,000 to 25,000 genes in their code (human beings have 25,000 genes; i.e. our bodies produce 25,000 different proteins). The number of genes does not necessarily imply intelligence of a life form (organism). The mice and the human beings have both approximately 25,000 genes (there are even plants with more than 25,000 genes).

Advances in technology have allowed reading the code in cells in the past decade or so[5]. Scientists have started sequencing (decoding) the genes in various life forms (organisms) such as some bacteria, plants and animals including humans. There are approximately 500 genes which have been found to be identical in all the organisms. This shows that certain codes which produce proteins have remained intact over the course of billions of years in all the life forms (immortal genes). These are all fundamental genes related to the basic processes (machinery) in the cells related to replication, protein making, etc. This common set of code (common genes) illustrates how life has relied on a unique set of basic cell functions (machinery) throughout the history of life regardless of organism.

Connectivity of Chemicals in Cells – Biological Networks

The complexity of a life form (organism) is not just the number of genes in its code but also the various interactions that take place in it. In other words, when cells produce chemicals, the resulting chemicals interact with each other which forms a huge network of interactions in the organism.

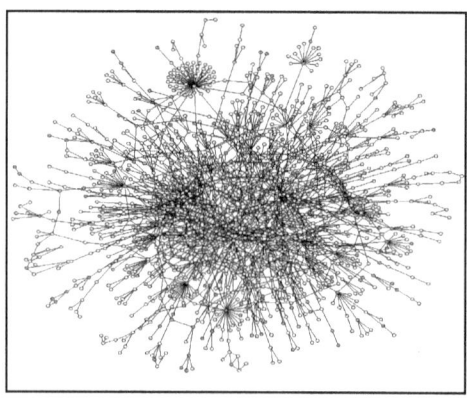

Figure 23 – Depiction of Complex Biological Networks

Therefore, understanding the code in the cells was necessary but not

---

[5] The infamous Human genome project's main goal was to read the code in the human cells.

sufficient to completely comprehend how an organism functioned. As we have seen earlier, the code in the cells generates thousands of chemicals (proteins) depending on the organism selected (in the case of humans this number was approximately 25,000). In other words, human body produces 25,000 different chemicals by reading the code in its cells and manufactures these chemicals.

However, the functions of life (i.e. metabolism, reproduction, etc.) are all carried out by various interactions among the proteins and other chemicals present in the cell (these other chemicals may be either present in the cell or may come from outside the cell through various transport mechanisms such as food digested by the organism may produce sugar that gets transported through the blood).

Metabolic Networks: One of the main characteristics of life is the metabolism which converts energy for the use of the cells and the organism in general. Metabolism entails interactions between various chemicals in the organism. Figure 24 is a simple metabolic network illustrating the concept of interactions between chemicals.

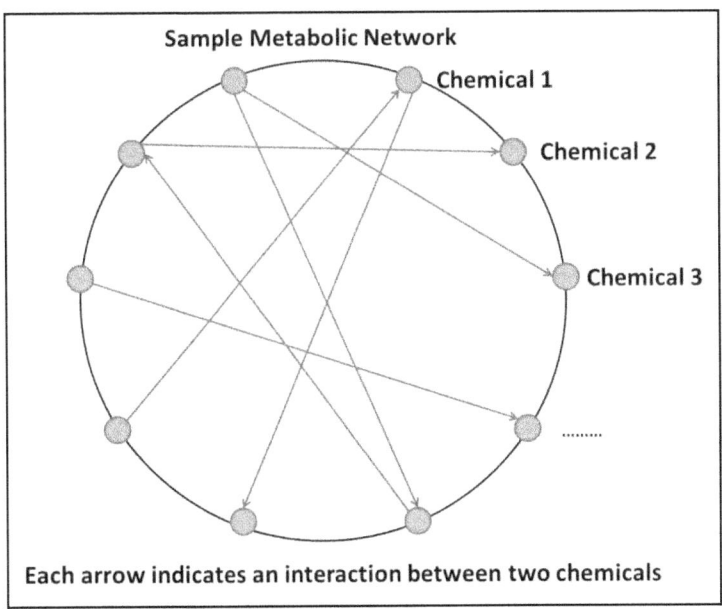

Figure 24 – Sample Metabolic Network

The following highly complex diagram is the human metabolism network which shows the interactions in the human body to give us energy which in turn helps in sustaining our lives. There are 2,626 different chemicals involved during these reactions in this network and there are a total of 7,439 chemical reactions that take place in the human body for metabolism[6].

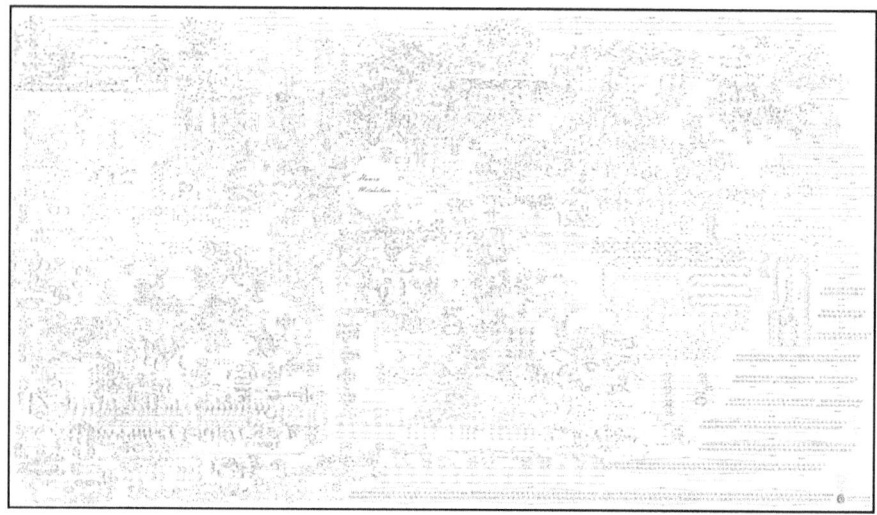

Figure 25 – Human Metabolism Network (Source: http://humanmetabolism.org)

Needless to say, it is a highly complex network and it is still being analyzed and the numbers being revised by the scientists. Given that human beings are quite advanced organisms and also appeared relatively late in the history of Earth, human metabolism reflects the advances and refinements that happened in the last few billion years of life's history on Earth. However, even a simple unicellular prokaryote like E.Coli bacterium has a fairly complex metabolism network as shown below.

---

[6] These figures are obtained from http://humanmetabolism.org and the site was accessed on August 18th, 2013.

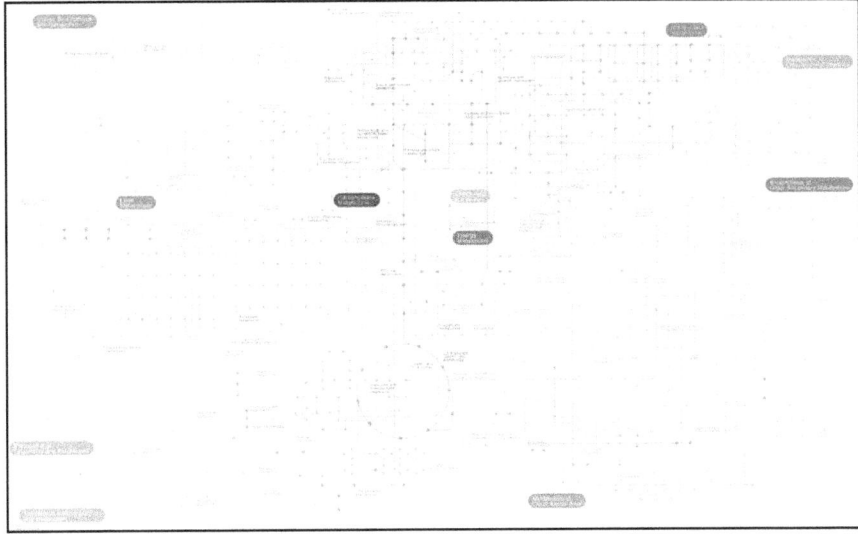

Figure 26 – E.Coli Bacterium Metabolism Network (*Image Courtesy of Kyoto Encyclopedia of Genes and Genomes – KEGG in GenomeNet website*)

Protein Networks: The proteins produced by genes (the code) in the cells interact with each other in the organism. Figure 27 is a simple protein network illustrating the concept of interactions between chemicals.

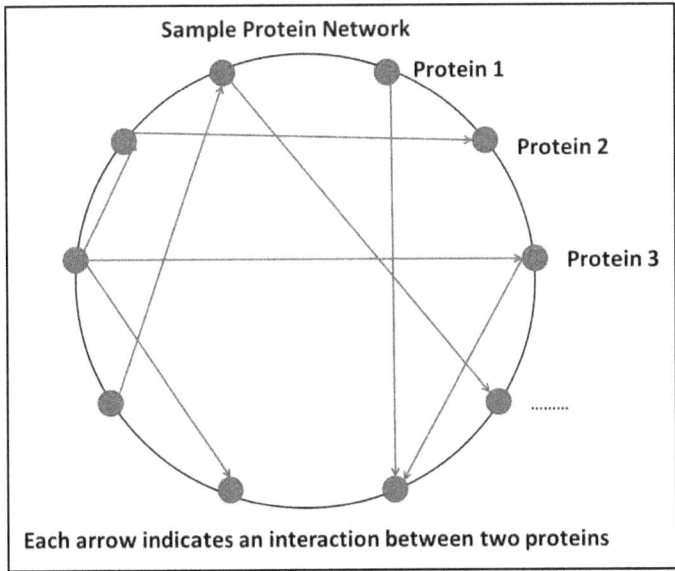

Figure 27 – Sample Protein Network

In fact life functions are carried out by such protein to protein interactions. These functions include cell signaling, growth and development, metabolism, etc. For example, there are approximately 25,000 different proteins produced by the code in the cells of the human beings. Correspondingly, there are hundreds of thousands of interactions taking place among these different proteins in the human body. The exact protein interactions are still being worked out.

Gene (Code) Regulatory Networks[7]: The genes (code) in the cells produce chemicals called proteins. However, there are inputs that determine whether a protein will be produced by a gene or not. Depending on the value of the inputs, protein may either be produced or not. Some inputs may promote (reinforce) the production of proteins while some others might inhibit (repress). Furthermore an input to a gene can be another gene which creates a network of dependencies among genes as illustrated in the below figure.

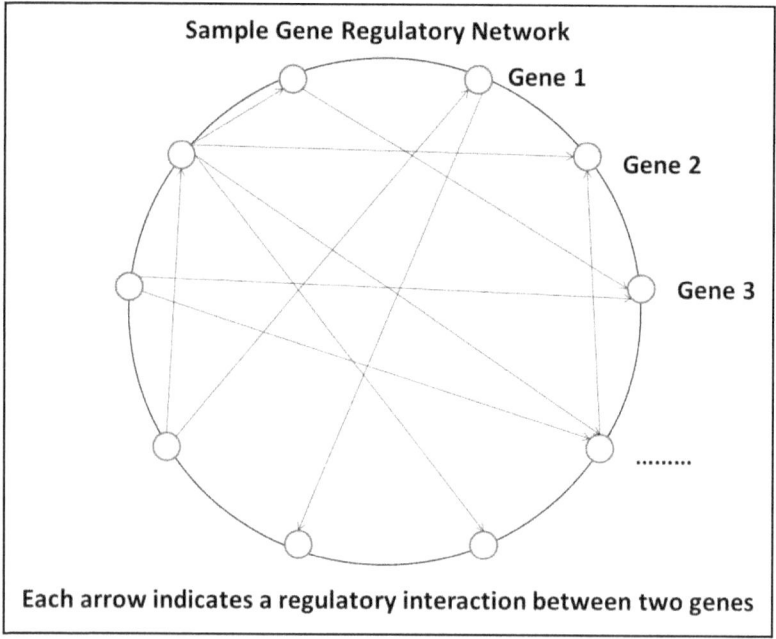

Figure 28 – Gene Regulatory Network

---

[7] This is called transcription regulatory networks in the scientific nomenclature.

The genes are indicated with circles. The arrows between the genes indicate that a gene acts as a regulatory input for another gene. Hence, an input gene affects the production of protein by another gene. In other words, genes provide feedback to each other which impact their production of proteins.

It is quite interesting to note that metabolic networks include various proteins produced by genes. Hence, the gene regulatory networks may affect the metabolic networks and the protein networks (in other words the biological networks are also interconnected (inter-network connectivity) at various points since a gene in the gene regulatory network produces a protein which gets used in one or more of the metabolic reactions in the metabolic network and also participates in protein to protein interactions. An example of inter-connectivity among biological networks is illustrated in the following figure.

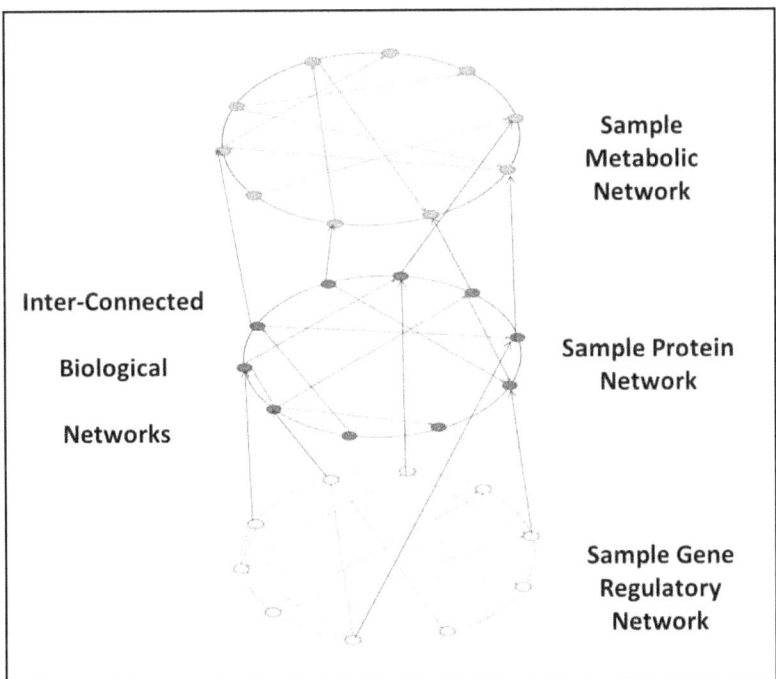

Figure 29 – Inter-Connected Biological Networks

Top circular nodes indicate chemicals in a metabolic network, middle circular nodes indicate proteins in a protein network and bottom circular nodes indicate genes in a gene regulatory network. Bottom circular nodes connected to middle circular nodes indicate the genes that create related proteins. Similarly, middle circular nodes connected to top circular nodes indicate the proteins that participate in metabolic networks as chemicals. Hence, biological networks have intra[8] and inter[9] connectivity aspects.

Signaling and Transduction in Organisms

Organisms have complex biological networks which are essential for carrying out their life processes. As life processes are carried out, there is extensive communication and signaling between the cells of an organism and even within the cells themselves. Energy or information can be converted from one form into another during this communication and signaling which is called transduction.

Our vision for example relies on converting optical energy (light arriving in our eyes) to be sent to brain through our nervous system so we can form the images in our brain for what we see. The same can be said for our five senses as humans. They are triggered by external stimuli and we convert these external signals (such as light, sound, heat, smell, etc.) to signals that our body can understand, transmit and process.

Our cell borders (membranes) contain special chemicals (called ion channels made up of proteins) that allow them to open or close their gates based on the signal that they receive through other chemical materials. In other words they generate and transmit electrical signals which allow the cell to sense its surrounding, receive information and make decisions and act on them.

For example there are certain chemical materials called hormones which are released by various organs in the body. They then go through the blood circulation system and finally reach the target cells which have the correct gates at their borders and they open and allow the hormones to enter into the cell. A response is then triggered within the cell depending on the hormone received.

Cells may also possess something called signal transduction cascades. That is, a recipient of a chemical signal in the cell can trigger one or more

---

[8] Intra refers to connections within a network.
[9] Inter refers to connections between the networks.

additional signals in a cascaded manner within the cell. During the cascade, signals can be amplified, converted from one form to another, etc.

Organisms rely on signaling and transductions heavily to perform their functions and to sustain their survival.

Adaptation of Organisms

Long Term Adaptations: The code (genes) in the cells of an organism has changed over time through addition of new genes, modification of genes or deletion of genes. The code in each cell is made up of hundreds of millions or billions of chemicals (depending on the organism) connected together like a chain. However a small number of chemicals in this long chain can randomly change due to certain effects (e.g. ultraviolet radiation, errors in cell division and replication, etc.). These changes referred to as mutations obviously result in changes in the genes for that organism and will in turn also cause changes in the proteins produced.

In such cases, altered genes will produce sometimes superior genes producing better functioning proteins, or new proteins capable of new functions; but in some cases they degrade the existing functions of existing proteins. In extreme cases, malfunctioning proteins as a result of mutation may even be lethal for the organism.

In other cases, the altered genes may supplement the existing genes with a new one which provide additional advantages (or disadvantages) for the organism. In such cases, the mutant (the altered individual) will have advantage (disadvantage) in the population over the other members of the same organism. Once a mutation occurs, chances are it is a matter of time for it to increase in the population due to its advantage or conversely to decrease due to its disadvantage. With each successive generation the ratio of advantageous gene to the whole population increases as well. Since organisms have been around for millions of years (or even billions for some organisms), there will be plentiful generations to inherit the advantage which will eventually penetrate the population in a dominant manner. The time effect is very important in gene alterations[10].

---

[10] The interested reader can look up population genetics for the diffusion of mutations in a population.

Another adaptation type is due to changes in biological networks discussed earlier. The biological networks described in the previous section differ from organism to organism. In other words each and every organism has different biological network structures. Each organism has built over time a different set of metabolic, protein and gene regulatory networks and interconnections among them. Furthermore, the networks for a given organism have also changed over the course of time due to millions or billions of years of adaptation and gene alterations.

Therefore the biological networks along with various chemicals that are involved in them define an organism and its internal interactions and functions. They are very important in understanding the life processes involved for an organism.

We have earlier seen that Earth has gone through major changes throughout its history. Those changes represented the external environmental changes for organisms living on Earth. Hence changes in organisms (in terms of altered genes and biological network connectivity) provided either advantage or disadvantage based on the external environment around them.

If the changes were such that it increased the overall success of the organism to sustain its survival rate in its environment than those changes diffused further in the population of the organism through replication. Those that reduced the overall success of the organism gradually vanished since other members lacking those changes had an advantage. Hence adaptations were either naturally promoted or filtered out depending on their fit to the external environment. Several species have become extinct throughout the history of Earth and some are still endangered due to their incapability adapt to external environment conditions (e.g. climate change).

Suppose an altered gene today produced a new protein that allowed human beings to survive easily and comfortably in freezing temperatures. And also suppose that Earth entered into a new ice age whereby Earth entirely became a snowball. Such an altered gene would allow the mutant human beings (i.e. those that have the altered gene) survive effortlessly and those who lack the altered gene would probably struggle or die out in such a harsh and cold environment. But if the Earth does not enter an ice age then the altered gene might not pose any advantages to the mutant human beings and the normal human beings lacking the altered gene might have an advantage and survive better in normal Earth conditions. So the adaptations in organisms provide mostly either advantage or disadvantage with respect to the external environment of the concerned organism.

Short Term Adaptations: The organisms respond to various stimuli in their environments. These are short term adaptations and do not cause permanent changes in the structure of cells, genes and the overall connectivity of biological networks. They are mostly transient and temporary responses and vary over time. We human beings for example have five traditional senses; namely our vision, hearing, tasting, touching and smelling. We behave; i.e. technically respond to stimuli based on these five senses. In fact most of our daily lives are preoccupied with things we perceive through our senses. We consciously do not operate our life processes which take place in our body (in all the cells) but rather we carry out activities which are mostly, if not completely, related to the responses to our perceptions acquired through our senses. Our bodies almost operate independent of our conscious intervention which is more geared towards short term adaptation in terms of responses to stimuli.

It is crucial to note that different organisms respond to different stimuli, in other words organisms have different senses. For example, many bacteria use quorum sensing to coordinate the production of proteins in their cells. Quorum sensing is a system of stimulus and response correlated to population density. Quorum sensing bacteria produce and release chemical signal molecules that increase in concentration naturally if more bacteria are present in an environment. If the bacteria detect a minimal threshold stimulatory concentration then they altogether produce a certain chemical. This is a very interesting response to stimuli since it allows bacteria to communicate with one another and to coordinate their production of chemicals, and therefore the behavior, of the entire community.

Other non-human sense examples include bats emitting and perceiving reflected sound to determine the location of objects, several species of fish sense the change in electric fields in their environs, several bird species sense their direction based on Earth's magnetic field.

Even though individual daily responses to stimuli do not change the genes, proteins or the biological networks in a permanent manner, it is important to mention that all sensory mechanisms have been established by an organism possibly millions or billions of years ago through genes producing proteins and / or related biological network changes. There is a wealth of biological information related to sensory perception of organisms and how they changed over time through changes in genes.

Connectivity Aspects of Biological Networks

One interesting aspect of biological networks is that their connectivity exhibits certain properties. More specifically, it has been shown that a relatively small number of nodes (dots) in biological networks have large number of connections. On the other hand a relatively large number of nodes (dots) in biological networks have small number of connections.

That is, a certain small number of chemicals in metabolic networks is affected by a large number of other chemicals; or a certain small number of genes in gene regulatory networks is affected by a large number of other genes and finally a certain small number of proteins in protein networks is affected by a large number of other proteins.

In other words, a certain small number of chemicals, genes and proteins have high connectivity in their biological networks. They tend to be quite important since their high connectivity indicates multiple functions dependent on them. Any changes in these highly connected chemicals, genes and proteins may significantly disrupt the functioning of the organism due to high dependencies associated with them. For example any altered gene which changes one of these highly connected chemicals, genes or proteins may render the network disconnected and dysfunctional for the organism itself. That is why these highly connected nodes (dots) increase the fragility or sensitivity of the organism in general. In other words, changes in these nodes (dots) may render the entire organism or a portion of it dysfunctional.

On the other hand, the less or poorly connected nodes (dots) have only a small number of connections to other nodes (dots). Hence, they increase the robustness of the organism or a portion of it since changes in these nodes may not cause significant disruptions or malfunctioning in the organism in general. On the other hand, by and large they may not be highly critical for the functioning of the entire organism or even a portion of it.

The Shape (Morphology) of an Organism

The entire shape and the body plan are coded within the cells of an organism. In other words, an organism's body shape unfolds from its embryo stage to maturity stage based on the code (instructions) available in its cells. The genes responsible for development, i.e. its body shape formation, have been discovered and are quite similar in several organisms. There are of course variations from organism to organism but these are essentially different values of certain parameters in the corresponding body shape formation genes.

It is quite interesting that the body shape design of an organism is available in every cell of the organism. It is embedded within each cell and preserved during the cell divisions or replications and gradually unfolds during the development of the organism.

Innovation in Life Forms (Organisms)

The unit of life is the cell. Changes in cell structure (as the hardware platform) such as prokaryote and eukaryote cell types are the first type of innovation and have enabled a multitude of new organisms throughout Earth's history.

The code changes in the cells (as the software) which result in changes in the production of chemicals (i.e. proteins) within the cells are the second type of innovation in organisms. Altered genes are such code changes which have occurred incessantly throughout the history of life and continue to do so as well.

The third innovation type is the changes in the connectivity of biological networks; i.e. the connections corresponding to interactions among chemicals, genes or proteins may vary over time which concomitantly alter the functions or life processes of an organism.

The characteristics of life point to a fundamental unity after the Life Bang. On the other hand, the three innovation types point to a fascinating source of diversity that operates on that fundamental unity.

These three innovations may potentially alter the body shape (morphology) and the various functions for an organism. They can potentially change all the life characteristics of an existing organism or even create a new organism.

The staggering variety of organisms (species) that we see around us in terms of microbial life forms, plants, animals, etc. are a testimony to these three innovations which happened over the course of life on Earth and it still continues to date. We owe the abundance of life around us to these three innovations.

Some Observations about Life on Earth

Most of what we human beings see as life is the macroscopic life consisting of plants and animals. However this is the minority, not the majority of life on Earth. Earth is teeming with life including both macroscopic and microscopic as well. There are organisms living in extreme environments successfully which would simply be inconceivable for plants and animals. There are life forms under ice caps, in the middle of hottest and harshest deserts, in deep oceans within the volcanic vents. There are bacteria living in clouds below freezing temperatures under constant bombardment from ultraviolet radiation. So life thrives even in unlikely environments.

All life on Earth is through complex chemicals based on carbon element. In fact, there are six main abundant elements which make up most of the chemicals that life uses; these are carbon, hydrogen, nitrogen, oxygen, phosphorus and sulfur. Of these six, only hydrogen formed recently after the Big Bang. The other five, including the carbon, had to be manufactured in the deep cores of stars millions or billions of years later. So these five additional elements were somehow present during the solar system formation (transported from other unknown stars' previous explosions).

Carbon is a very versatile element which provides sufficient flexibility and stability for combining with other elements to form various chemicals supporting life processes. It exists in the code in the cells (i.e. genes), in proteins and in almost all chemicals involved in metabolism. In the temperature ranges common on Earth and also in water, carbon exhibits outstanding properties as a building block for life. Hence carbon based life has thrived on Earth. Silicon might be an alternative element to carbon and already exists in some organisms such as plants and to a lesser extent in animals. Under general conditions on Earth carbon is superior to silicon. However, silicon would be better than carbon under considerably higher and lower temperatures or in solvents other than water in terms of stability and flexibility.

Life on Earth requires a liquid medium. Water, with all its advantages with respect to carbon based chemicals, has been the suitable liquid medium for life processes under normal Earth temperatures and pressures.

The chemical structure of the code in the cells of all life forms (organisms) is the same and it is referred to as DNA (deoxyribonucleic acid). In other words, no natural life form as of yet has been encountered which possesses the characteristics of life and is not based on DNA. Interestingly enough, scientists in 2012 have synthesized a new set of molecules called XNA (Xenonucleic acid) which possess similar properties as the DNA. That is they can store genetic information and also evolve just like the DNA. These synthetic molecules can potentially form a new springboard for novel forms of life.

So the code (the software or the instructions) in cells of all current natural life forms relies on the same chemical structure called the DNA. The code in the cells is a one-dimensional code stored chemically as a double helix and produces 3-dimensional proteins eventually with specific structures and functions.

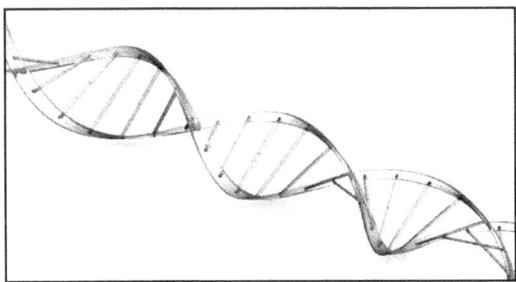

Figure 30 - DNA Picture

The significance of carbon, water and DNA to life on Earth might be coincidental or might reflect dependency on the environmental conditions on Earth. So far, we are familiar with only one planet, our Earth, as a life form but we have seen that potentially there is a staggering number of planets in the universe as a rough estimate. Knowing a single planet would not allow us to assess all different feasible life forms in other planets. There could be planets where life is based on another element than carbon (e.g.

silicon, others, etc.), another solvent than water (e.g. methane, ammonia, etc.), and another chemical structure than DNA for coding life. Other planets might be smaller or larger than Earth and might be very close or very far away from its star. Hence, conditions on such planets might be very different from our Earth in terms of temperature, pressure, existence of atmosphere and oceans. In other words, the overall landscape for life might be alien or completely unrelated to the landscape on Earth as we know it. Yet, life could still emerge and thrive under such alien conditions and might be substantially different from ours and would adapt to the conditions present in those planets.

Life on Earth has taken almost a billion years to emerge and has existed for two to three billion years with microorganisms. The macroscopic life has emerged mostly in the last billion years and has substantially diversified after the Cambrian explosion in almost the last half a billion years. It is still not clear whether such prolonged timelines are essential for life on other planets or not. Certain stars have much shorter life times and their planets may not accommodate life if such protracted timeframes are indeed essential for life. If that is the case, life would be restricted to stellar systems with a minimum lifetime of billions of years.

We can hypothesize that life would still have a parts list (i.e. building blocks as part of toolbox) and same or similar innovation types would prevail in different life forms (organisms) on other planets. We have yet to find it out but we cannot preclude the existence of life on other planets. Given the astonishing number of potential planets in the universe, life not emerging in anyone of them is statistically a dubious possibility at best.

The Life Bang after Big Bang

The Life Bang subsequent to Big Bang has logically partitioned the universe once and for all into two main categories of matter; the living and the non-living. The living matter possesses the characteristics of life. For the first time in the universe, the living matter would have the capability to respond differently from the non-living (inanimate) matter. More specifically, the living matter has an autonomous capability to decide for its responses, whereas the non-living matter would duly obey the laws of nature with no intrinsic cognitive response capability.

To illustrate this let's think of two scenarios. In the first scenario let's assume you are holding a living bird in your hand and in the second scenario you are holding a stone in your hand. If you open your hand and let go of both in the two scenarios the results will be different. In fact, we cannot in most cases guess which direction the bird will fly but we can exactly guess that the stone will fall on the ground under the effect of gravitational force). The bird can by itself decide which path to take and furthermore if you repeat the same experiment many times with the same bird you may get different results every time.

Life Bang rendered the universe (at least the portions where life exists) precarious compared to Big Bang and also brought about further innovation potential as we will explore in the next chapter.

Further Innovation at Planet Scale – The Technology Bang

Big Bang followed by Life Bang has created diverse organisms on Earth. Big Bang along with macro scale innovation (i.e. star deaths) had provided the innovation building blocks (our parts list in our toolbox) on our Earth. Earth was bestowed with more than 100 chemical elements. Furthermore, the laws of nature especially the electromagnetic force combined some of these elements to form composite matter such as chemical compounds enriching the building blocks (the parts list in our toolbox) on Earth. These were natural purposeless innovation schemes via straightforward application of the laws of nature.

However, Life Bang changed this gradually, nonetheless considerably. The sensory mechanisms in living organisms and their short term adaptations allowed distinct responses from living matter which are very different from their non-living counterparts. For example, sensory mechanisms of some organisms identified chemical nutrients which led them to food sources. In some cases, hazardous situations were avoided by sensing toxic matter or other organisms that would pose danger to them. Such sensory mechanisms permitted them to avert dangerous and lethal outcomes. But the most important consequence was that living matter behaved different from the non-living matter. The outcomes encountered with the inclusion of living matter were not explicable just by the blind application of the fundamental four forces of nature as envisaged for the non-living matter.

The adaptation and the autonomy inherent in the living matter brought about one major innovation which we will refer to as Technology Bang. In this book, we will define technology as the "…making, modification, usage, and knowledge of tools, machines, techniques, crafts, systems, and methods of organization, in order to solve a problem, improve a preexisting solution to a problem, achieve a goal, handle an applied input/output relation or perform a specific function"[11].

Two important remarks are worthwhile to mention here. The first is that there is an implicit inference (reasoning) implicit in the making, modification, usage, and knowledge of tools, machines, techniques, crafts, systems, and methods of organization. This would be unattainable by the non-living matter. The second is that there is a purpose in technology stated as solving a problem, improving a preexisting solution to a problem, achieving a goal, handling an applied input/output relation or performing a specific function,

We can deduce that technology is the means to an end; it accomplishes a purpose. The universe was transformed once again with the Technology Bang. A new innovation type had emerged which as we will see would transform at least the Earth substantially. Technology Bang introduced purposeful innovation which was very different from the natural purposeless innovation. Technology as an innovation type is exclusive merely to the living matter. The non-living matter would not be able to utilize technology as an innovation type since it lacked inference making and purpose formation capabilities. So Life bang was a pre-requisite for the Technology Bang.

In our daily lives, we ordinarily use the term technology to represent various tools[12] used by human beings. However, technology has been created and used by many life forms (organisms) other than humans.

Primates such as monkeys, orangutans, chimpanzees are known to use tools in the wild. Some of them are known to crack shelled food such as nuts with stones. Some monkeys are known to pluck out hair from human beings and then floss their teeth with it. Some primates use sticks to gauge the water depth and to kill ants to eat them. I myself have observed a crow on the beach trying to crack a nut first by its beak, then by stone while it is on the ground; when it could not crack the nut it flew almost a meter high

---

[11] Source: Wikipedia - http://en.wikipedia.org/wiki/Technology; Accessed on August 20th, 2013
[12] A tool is any physical item that can be used to achieve a goal. Hence pretty much everything around us that we use is a tool. Examples are furniture, stationary, cars and any other transportation vehicles, electronic devices, etc. In this book, we will exclude living beings but include all the rest under the tools concept.

with the stone held by its feet and then released the stone from the air down which finally cracked the nut. Nest making by birds is another example for the usage of tools. These examples can be enriched with several other ones by observing the animals in their natural environments.

In this book, we will focus on the technologies created and used by human beings since the technologies created by human beings have to a very large extent surpassed other organisms in terms of variety and complexity on Earth. We as human beings have taken it to a level that technology has become embedded in our lives. As I write this book in our living room, I realize that everything has been made or manufactured to meet a purpose and obviously my wife and I have acquired them for a reason. I am using a computer to type the manuscript of this book while sitting on a sofa facing the TV and the music system across from the dinner table surrounded by chairs (all these items are artifacts of technology). I look at the entire room and the only natural non-manufactured item is the plants in the room (and myself). At first glance, it is easy to notice that the number of distinct items in the room is much more than one hundred and it is ironic to have merely a handful of natural items which are the plants (and myself). All the rest are the results of the Technology Bang serving a specific purpose. With some exceptions this may be true for a good majority of us (just look around you to realize the extent of this). We have become so accustomed to technology produced items that we take it completely for granted.

I sometimes take a short trip to visit the diverse places in and outside Dubai. It provides a unique blend of sharp contrasts in terms of the desolate dessert, turquoise waters of Arabic Gulf and a highly urbanized city. In some of these trips, I visit the Bedouin villages in the dessert leading relatively primitive lives far out in secluded places (seemingly much simpler yet much happier lives compared to urban lifestyle). These visits make me shockingly realize how technology bound we all have become in our lives. The small number of items in Bedouin dessert tents reflect their simple lifestyle, possibly not more than tens of items with half or more made of natural materials, compared to thousands in our houses which are almost exclusively manufactured products of technology.

The human history has witnessed the use of technology from early civilizations onward. Initial use of stone tools can be seen more than millions of years ago in Africa. These early stone tools (early artifacts of technology) might have been used for hammering, chopping, digging and butchering among others. Additionally, early humans might have used tools made of wood and hard-body parts of animals. Later on, axes and other large cutting tools were added to their toolsets which was followed by spears, darts, arrows, etc. Note that some of these tools are combinations of multiple parts to achieve a purpose. (a spear is a point attached to a shaft for making a projectile weapon). In parallel, early humans also extended their toolbox by adding new materials such as bone and ivory.

History and archeology museums are interesting places to visit in order to understand the advancement of ancient civilizations and their progression over time. Especially the tools that they used provide great insights for the lifestyles that they led and their cognition capabilities. The tools were used for hunting, fishing, dwelling, and also for some of the other activities in their daily lives.

So the technology artifacts that humans have produced were becoming more advanced over time in terms of the number of parts, the materials used and the purposes accomplished. As they say, necessity is the mother of all inventions. Humans have developed numerous tools to meet their necessary needs. In some cases similar tools were developed recurrently (almost like reinventing the wheel) or were learnt from the craftsmen nearby in an informal manner. Eventually, one important change has helped to revolutionize technology drastically. The division of labor and specialization of cooperating individuals in the way technologies were created and used introduced radical changes. It has triggered the growth of economies through trade, industrialized production and consumption processes, and has led to the rise of capitalism and entrepreneurship. Starting from 18th century new technologies were developed which once and for all triggered a revolution commonly referred to as the Industrial Revolution.

Water and steam were used to mechanically spin cotton driving the productivity much higher in the textile industry. Steam engines rapidly expanded for industrial use and transportation. Iron production became easier thanks to advances in related technologies. Coal replaced wood and other bio-fuels since coal mining was more efficient. Several chemicals and metals were used by different industries. This created demand for metal parts and making machine tools. Machines were used in agriculture and in manufacturing glass, paper, transportation products and many others. Steam, electric and finally gasoline powered cars were invented.

In fact the rest was history as they say. The mechanical, chemical (including biological) and electrical (including electronic) technologies among others have shaped our lives through the use of targeted tools.

Observations about Technology

Just like we have seen in the case of Life Bang for life forms (organisms), a staggering amount of diversity has occurred in the technologies and the actual related tools we use. Technology Bang has created an astonishing variety of tools with different colors, shapes, materials, components and their assemblies, as well as their purpose.

Technology tools essentially apply the laws of nature (science) to meet a purpose. Developments in physics, chemistry, biology and engineering have shaped the technology tools that we use. The underlying principles of steam engine, laser, telephone, telegraph, radar, computers, mobile phones and many other tools that we use strictly reside in science and engineering[13]. Hence advances in our scientific knowledge have concomitantly inspired and catalyzed advances in technology related tools.

Figure 31 – The World's First Production Car (Image Courtesy Mercedes Benz)

---

[13] Even engineering can be reduced to application of science by means of detailed analysis.

Above is the world's first large scale production car, the Benz Velo of 1894 with a top speed of 20km/hour (i.e. more than 13 miles/hour), an engine size of 1 liter and 1.5 horsepower (literally the leading edge car of its time). If we compare it to a leading edge car of today, we would see similarities as well as stark differences. In fact at first glance both seem to have the basic components, namely the power generation, power transmission, suspension, steering, braking, and the comfort components like the seating. However, the details vary substantially in terms of how these basic components are implemented. Advances in technology have penetrated into these basic components and have altered their implementations over time. In some cases, mechanical components have been partially or completely replaced by electrically controlled systems, Comfort components today include heating, air conditioning, fairly advanced dashboard indicators, GPS, audio and video unlike the first generation cars. Obviously, the car has progressed extensively over time and the progression is conspicuous in all its basic components.

As a matter of fact, the car example can be extended to other technology tools that we use. Namely, we can analyze other technology tools that we use by breaking it into its basic components and analyze the individual components, their interactions and the technological advances within the components themselves. The white and the brown goods we use at home, furniture, kitchen utensils, electrical machines, computers, mobile phones, etc. have all progressed and changed over time and continue to do so as well.

The technology tools naturally serve a purpose. We have also broken down the technology tools into its basic components such as power generation, power transmission, suspension, steering, braking, and the comfort in the car example. It is quite evident that the basic components of a technology tool has certain functionality. The combined functionalities of the basic components deliver the intended purpose in the first place from the technology tool. Needless to say, the intended purpose for a technology tool is in most cases very different from the individual component functionalities. For example, the intended purpose of a car might be transportation from point A to point B in a safe and comfortable manner whereas the function of one of the basic components might be to generate power and the other to transmit power (neither of which is the ultimate intended purpose of the car). So the components themselves are the means to an end; they all contribute to achieve the intended purpose of the technology tool.

Modeling Technology Tools

Technology tools can be modeled as a combination of components connected together to achieve one or more purposes.

Naturally the components are not connected randomly, but instead in a specific configuration. Not every component is connected to every other component but rather each component is connected to specific one or more of the other components.

Each component has a certain well-defined functionality and contributes to the overall technology tool.

Each component is made up of one or more parts. The parts that make up a component interact with each other.

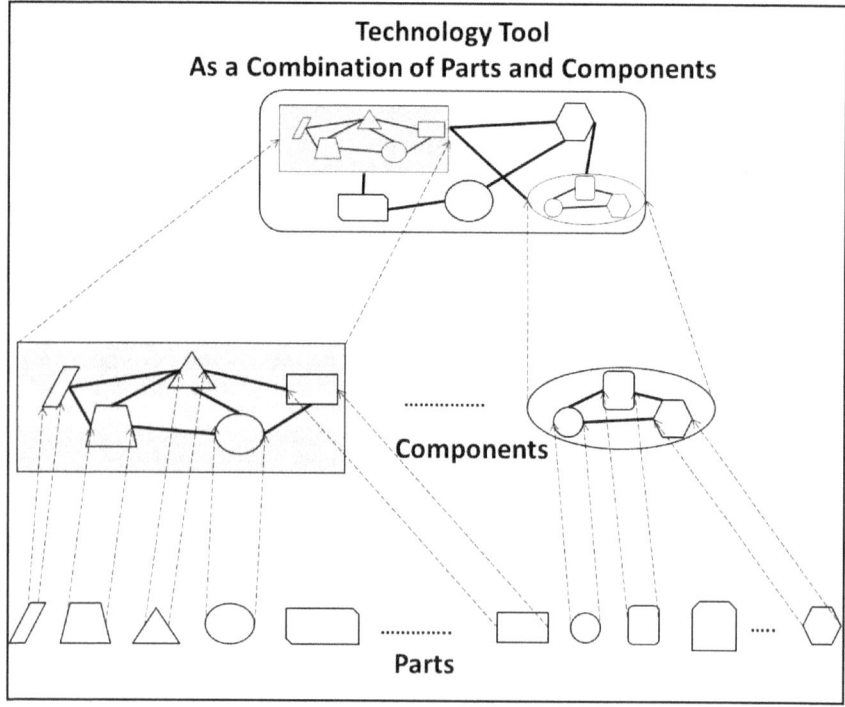

Figure 32 – Hierarchy of a Technology Tool

Holistically, the entire technology tool can be thought to be comprised of components and parts as well. In fact, it can be modeled hierarchically as a combination of components at the first level and as a combination of parts at the second level.

Interestingly, this hierarchy can be extended several times. Each part at any level can be thought of as a component which is further made up of parts.

Innovation of a technology tool can be thought of as delivering the intended purpose through the combination of components and parts at multiple levels. In real life, the technology tools can be designed with not only the intended purpose in mind but also certain desired features and performance characteristics may be specified. These desired features and performance characteristics directly and/or indirectly impact the components, their connections among each other and also the design of the components themselves including the parts in them.

For example, a simple coffee mug can be specified to withstand a range of pressure and force values in order to increase its durability. Such specifications may impose constraints on the materials used for producing the mug. Other examples might be design specifications in terms of shape, aesthetic factors, quality, safety, high performance under certain stringent conditions, cost, etc. Such factors and others will affect the technology tool design and / or production processes associated with it. In real life, technology tool design may go through iterations until simultaneously the intended purpose and the desired features and performance characteristics are achieved.

The general principles stated in this section can be applied to decompose any technology tool into its individual components / parts and their connections among each other. This analysis technique can be applied to any existing technology tool to understand its design (almost like reverse engineering).

Conversely, technology tool design involves synthesis of new functionalities from existing components and parts.

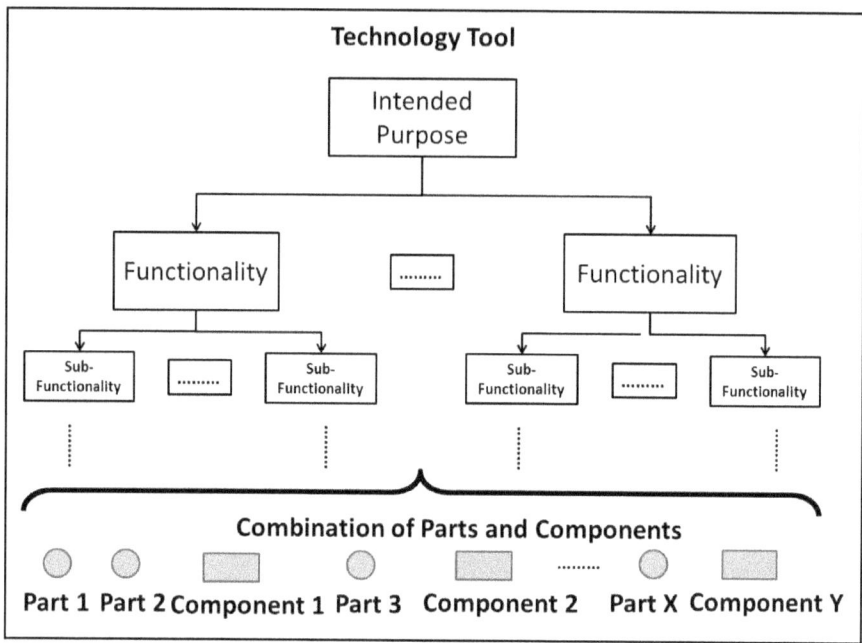

Figure 33 – Technology Tool as a Combination of Functionalities for an Intended Purpose

Figure 33 indicates that parts and components can be combined to form sub-functionalities as well as functionalities. Sub-functionalities can be combined to form even higher level sub-functionalities pointing to highly hierarchical nature of possible combinations in technology. This is quite interesting because technology tools are created from existing parts and components combined in novel ways. In a way, technology creates novel technologies from existing ones. Hence, innovation depends on novel combinations of existing parts and components arranged to provide new sub-functionalities and functionalities.

A flat wooden rectangular surface can be attached four legs in its corners and can be turned into a table. For those who assemble their furniture at home would definitely appreciate this concept of parts and components combination. Tens or sometimes hundreds of parts turn into a usable furniture piece such as a wardrobe, bed or sofa with a highly different purpose than its individual components and parts. The same can be said about other electrical and mechanical technology tools we use. When we look "under the hood" of devices, equipments and vehicles we use, we always perceive a similar situation; a group of components and parts combined together to form the final product. This concept also applies to information based products such as software applications which are designed as combinations of pieces of software codes.

Parts and Components in Technology Tools

Human beings have developed simple parts and components with specific functionality which when combined through complex connections may yield new technology tools. They can be thought of as the simple building blocks of more complex parts and components. It is analogous to aggregating or assembling a set of parts and components.

It is interesting to note that parts and components have developed under certain main categories, which we will identify as domains. The domains can be mechanical, electrical, optical, chemical, etc.

Mechanical components and parts exhibit desirable functionalities under the influence of forces or displacements. Examples of mechanical parts and components are wheels, shafts (a rotating part used to transmit power or motion), axles (a non-rotating part used to support other parts and components), wedges, screws, threads, joints, washers, joints, bolts, springs, bearings, gears, belts, cams, clutches and brakes. Needless to say each part and component can have many different types (a screw can be normal, threaded, etc., a gear can be helical, worm, etc.) and they can be made up of different materials impacting their strength, deflection and stiffness under different load and stress conditions. These components can be connected in several ways with the help of pins, fasteners, welding, brazing, soldering and gluing among others to provide additional functionality. For example transmission of power in a car can be obtained by combining shaft, bearings, gears, belt pulleys, etc. Design specifications would determine the actual parts and components to be used in terms of materials used and connections.

Mechanical technology tools are abundant all around us. The cutlery in the kitchen, mechanical tools in your storage room (hammer, screwdriver, etc.), most of the furniture at home, the mechanical parts of the cars, bicycles are all mechanical technology tool examples. The materials which are used in parts and components usually have certain finishing applied to them; for example it is quite common to see metal parts having paint, lacquer, plastic and oil finishes and wooden parts having varnish, paint, oil and stain finishes after sanding them.

Mechanical and civil engineering immensely take advantage of mechanical parts and components.

Electrical components and parts on the other hand exhibit desirable functionalities by harnessing the power of electrical energy (including electronics) in the form of currents and voltages. Examples of electrical parts and components are electrical wires, batteries, resistors, switches (relays), capacitors, inductors, diodes, transistors. These can be combined to form components such as integrated circuits which are more complex technology tools.

Figure 34 – Sample Integrated Circuit

Electronic circuits today include chips and other electronic parts and components which deliver a certain desired functionality by combining simpler electrical parts and components.

Figure 35 – Sample Electronic Circuit

Electrical components and parts can be combined to achieve new functionalities to implement logical circuits. Transistors, resistors, capacitors, etc. can be combined to achieve circuits that can perform logical AND and OR operations. That is, a threshold output voltage will be obtained if both inputs are at a threshold voltage value (AND); or a threshold output voltage will be obtained if one of the inputs are at a threshold voltage value (OR). Similarly we can further combine these new functionalities with other electrical parts and components to obtain even newer ones. This is the hierarchical nature of technology which we discussed earlier.

One of the chips used in earlier computers was introduced in 1971 and included 2,300 transistors in an area of 12 mm² which was revolutionary for its time. In 2012, about five billion transistors have been included in an area of more than 500mm² [14]. The difference between the two chips is a testimony to the significant advances in electronic circuits. Though the advances have been incremental and gradual over the years, enormous computing power has been squeezed in miniature spaces.

---

[14] Source: http://en.wikipedia.org/wiki/Transistor_count

There are ample examples of electrical technology tools around us. Audio and video systems at home, computers, mobile phones, electrical appliances that we use in general are all examples of electrical technology tools.

Optical components and parts on the other hand exhibit desirable functionalities by benefiting from the behavior and properties of light (both visible and invisible). Examples of optical parts and components are optical lenses, mirrors, diffusers, filters, polarizers, prisms, light splitters, fiber optics, lasers, etc. These can be combined to form components o create more complex optical technology tools.

Optical technology tools are used in lighting products, cameras, telescopes, microscopes, telecommunications, and imaging products among others. Fiber optic cables are commonly used in long distance high speed and high bandwidth communications (plays a critical role in carrying the long haul backbone Internet traffic). It is mind boggling to imagine that a single strand of fiber optic cable is able to carry the entire world's voice traffic for telephone communications. Similarly, a laser can focus and direct a light in a certain direction with a wide range of intensity levels.

Mechanical, electrical and optical technologies allow an impressive set of functionalities which technology tool designers can capitalize on based on their purpose and specific applications. Selecting the right parts and components, their connections and the right materials present a designer with an enormous flexibility and capability in order to achieve an intended purpose.

These technology examples can be extended to other domains such as chemical, etc. Each domain contributes with its own set of parts and components to the technology tool designer's potential toolbox.

Modeling Innovation in Technology Tools

Technology tools that humans use have to a large extent progressed over time. Innovation in technology tools have two types: the first one is the changes in an existing technology tool and the second one is the invention of new technology tools. The former can keep the purpose intact or even extend the purpose of the tool while making changes in the parts / components of the technology tool.

For example, television allows us to watch broadcast programmes and the cathode ray tube technologies have been replaced with LED and Plasma technologies recently (purpose remained intact but internal components have changed).

On the other hand there are televisions with internet access capability (the purpose of watching broadcast programmes has been extended with internet access and new components have been added along with technological changes in existing components).

The second type is the invention of new technology tools which did not exist before. This type fundamentally creates a new technology tool in order to meet a new purpose, not to enhance an existing one. This second type of advance is more infrequent compared to the first type. Let's say we create a new robot that performs as our personal driver. It can converse with us and drive us to our work place in a highly safe manner. This would be a new technology tool since it does not exist today.

Since the unleashing of Industrial Revolution, the parts and components of various technology tools have had incremental as well as abrupt changes. New materials have been discovered and invented which had superior properties to already existing ones. Semiconductors enabled transistors which subsequently opened a new era in electrical parts and components leading to an unprecedented variety of electronic components which continues to date. Polymers have enabled new versatile materials such as plastics and fibers. Polymers are used in several technology tools such as electrical appliances, paints, radio and television cabinets, coatings, adhesives and many other countless objects. Some polymers can withstand very high temperatures and have been used in spacecraft linings. While writing this book, a new material called Carbyne has been discovered which is forecasted to have higher strength and stiffness than any other known material (formerly it was graphene) so there is continuous development for new parts and components stemming from scientific discoveries and inventions. Nanomaterials are another example of recent material advances. The nanoscale dimensions (extremely small) of these materials render them suitable for very specific applications. They are used either by themselves or along with other normal materials like polymers to enhance their various properties such as stiffness, strength and their overall performance under various operating conditions. Nanomaterials pose a strong potential to significantly alter and enhance various materials in mechanical, electrical, optical technology tools among others and can be applied in various science and engineering fields to solve specific problems. Material advances have allowed humans to raise the bar in terms of performance and robustness in technology tools. Demanding technology tools which operate under stringent temperature, pressure, acidity, and other challenging environmental conditions have been rendered feasible through incessant advances in materials (e.g. deep earth and ocean applications, air and space applications).

Innovation in technology tools also transpire due to new combinations and connections of parts and components. If we look at a car, especially under the hood, from early 1900s and a recent one in 2010s we will see a much higher complexity in the latter one. There are many more parts involved and the connections among parts are also highly complex. The higher number of connections tends to bring higher performance and robustness to technology tools but sometimes also renders them fragile since it tends to increase the probability of failures as well.

Another innovation mechanism is to use an existing part or component as a new platform for various novel technology tools. They catalyze the design of new technology tools and act as a platform to build upon. Transport infrastructure (e.g. automobile and rail roads), is a platform technology tool which enabled vehicles (cars, trains, etc.). In information technology industry, certain computer hardware (e.g. Intel based computers, Macintosh computers, tablets, etc.) can act as a platform for an operating system (e.g. Microsoft Windows). An operating system can act as a platform that enables multiple software applications to run on it. Automobile platforms enable several new models of cars to be released by utilizing a common platform (the platform may consist of a single powertrain, chassis, etc.). A platform enables modular production (manufacturing) of technology tools by leveraging on common parts and components at the platform level while supplementing it with additional ones yielding further functionality and also novel intended purpose(s).

Transduction and Hybrid Technology Tools

We have so far discussed technology tools by segregating it into separate domains such as mechanical, electrical and optical. However, the natural progression of technology has blended these domains over time forming hybrid technology tools. Hence technology tools have become even more complex utilizing different domains based on the functionality required from different parts and components.

In applications where complex decisions and control are required, electrical domain can be more appropriate through the implementation of digital logic through sophisticated integrated circuits. In applications where high bandwidth and high speed information transmission is required over long distances, optical domain can be more appropriate (in some cases electrical domain can also be a viable alternative). In applications where large size tangible materials and their aggregations and assemblies are required, mechanical domain will be more suitable (e.g. large construction objects such as houses, buildings, bridges, furniture, etc.).

Hybrid technology tools require integration of these domains through the connection of corresponding parts and components. However, technology tool designers are confronted with the challenge of converting one domain's output to the input of another domain which are intrinsically incompatible. In such cross-domain integrations, special parts and components called transducers are required. Transducers convert a domain's output into another domain's input. They can be either sensors or actuators depending on their functionality. Sensors detect one domain's output parameter value and convert it into an input in another domain. Actuators obtain an input from one domain's parameter and convert it into an output signal (action) in another domain. For example a pressure sensor detects pressure value (in a mechanical system or material) and converts it into an electrical signal (current or voltage which feeds into an electrical circuit for further processing). A loudspeaker on the other hand is an actuator which takes an electrical signal (current or voltage) as an input and converts it into a mechanical motion or action (the mechanical movement of woofers and tweeters producing sound). Today's technology tools commonly operate in multiple domains (hybrid approach) by making use of transducers wherever needed.

Therefore usage of hybrid technology tools via transduction has further expanded the set of possible technology tools enabling either achievement of new purposes or achievement of existing purposes in an easier and more efficient manner.

Today's airplanes for example make use of all three domains. It is composed of thousands of mechanical components making up its fuselage, wings, etc. They also use fly-by-wire systems which transduce several measurements (e.g. altitude, position, speed, acceleration, wind, etc.) through sensors and convert them into electrical signals which are fed into highly complex flight control systems which are almost exclusively implemented in electrical domain. Once the commands are generated through the electrical control system, they are then transduced (converted) through the actuators into real action via the engine controls, flaps on the wings, directional changes in rudder, etc. which are in mechanical domain. The laser and the lighting system of the airplanes are examples of optical domain parts and components. So the airplane is a hybrid technology tool.

A digital camera on a tripod is another example of hybrid technology tool. The tripod is a mechanical component (composed of three parts). Taking the picture involves optical parts and components (sometimes highly sophisticated lens systems). The photograph is then converted from optical domain to electrical domain and stored as a digital file by the cameras. So the digital camera on a tripod is a hybrid technology tool.

When I took my car for its first 5,000km service in 2006, I remember the technician saying that they need to upgrade the entire software on my car to reduce the slight vertical wobbling in it. Even though cars seem as mostly mechanical devices (except the battery), their control systems for navigation, suspension, cruise control, etc. are being handled through electrical parts and components. And one can notice several transduction parts and components in them (sensors and actuators). And of course the lighting systems in the cars are optical.

There are also examples of hybrid technologies comprised of two domains. Electromechanical and mechatronic systems are examples of these whereby mechanical and electrical or electronic domains are combined during the technology tool design by means of transducers.

We have talked about the Technology Bang in this chapter and tried to formalize it to a certain extent. As an avid supporter of theoretical concepts with practical examples, the fascinating book "The Toaster Project"[15] talks about a person (Thomas Thwaites), making a toaster from scratch. It takes him nine months, travelling 1,900 miles in UK and costs £1,187.54 to build a self-made toaster from scratch which costs £3.94 at a nearby store. He disassembles the toaster he bought in the store and realizes that it is made up of 157 separate parts (some of those parts are even further made up of other constituent parts). The captivating part of the book is that he decides to build the toaster from its very basic raw materials such as steel, mica, plastic, copper and nickel. He tries to obtain or extract them from the earth's crust through short mining trips and then processes them in a primitive manner through his own efforts. The entire book demonstrates how accustomed we have become to using technology tools without realizing the astounding efforts and processes required in producing them (even their environmental implications). It also shows how division of labor and specialization have enhanced the effectiveness and efficiency of technology tools.

Cumulative Innovation in Technology Tools

Progress in science and technology have greatly diffused into the society and in turn impacted the technology tools we use. One important enabler for this has been the gradual and cumulative build-up of the science and technology knowledge in the form of pure theoretical as well as applied craftsmanship. Humans are, to the best of our knowledge, the only species that have developed systematic information and knowledge storage and dissemination systems in a structured manner. After the invention of publishing and more recently electronic information storage and transmission, the knowledge contributions have been cumulatively preserved and applied in practice. Our education systems, access to knowledge in the form of published and electronic materials, sources, etc. expedite our acquisition of knowledge. Needless to say, we all specialize in certain areas but collectively we possess a staggering amount of cumulative knowledge built over time in human history. Each generation has built on the previous one and has advanced the collective human knowledge. This has been a major enabler for the advances in technology tools.

---

[15] "The Toaster Project", by Thomas Thwaites, Princeton Architectural Press · New York

Today, we collectively possess a good top-down and bottom-up understanding of the technology tools we use. For any given technology tool that we use, we know its purpose, its structure, the components and the parts that it is comprised of and their connections and the materials used in it (to a minimum there is one or more persons who possess this knowledge in the world for any given technology tool). In certain cases, we have established legal barriers to preserve the knowledge behind technology tools in the form of intellectual property rights. However, even in those cases the legal entities or persons that own the intellectual property right are entitled to the knowledge behind those specific technology tools.

Connectivity Aspects of Technology Tools

The parts and components are connected in a specific manner based on their functionality to achieve the intended purpose of a technology tool.

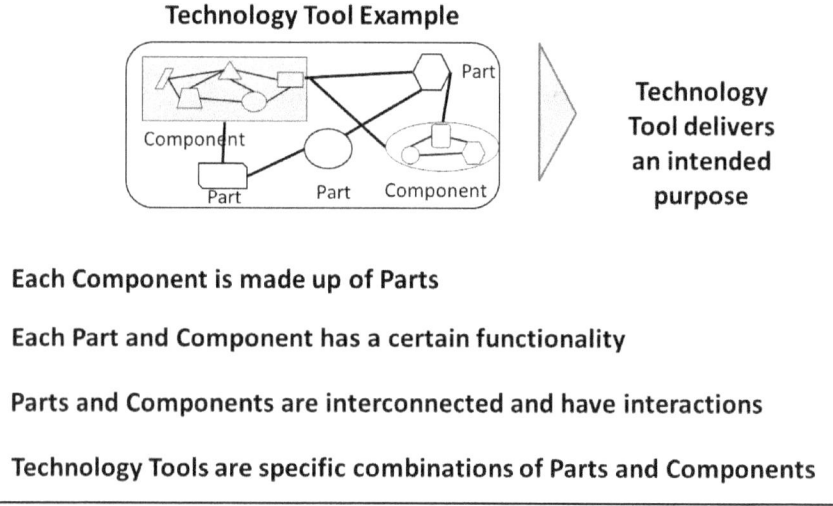

Figure 36 – Connectivity of Technology Tools

The connectivity of the parts and components can alternatively be represented as a network which is analogous to the biological networks we have seen in the previous Chapter.

UNIVERSAL INNOVATION

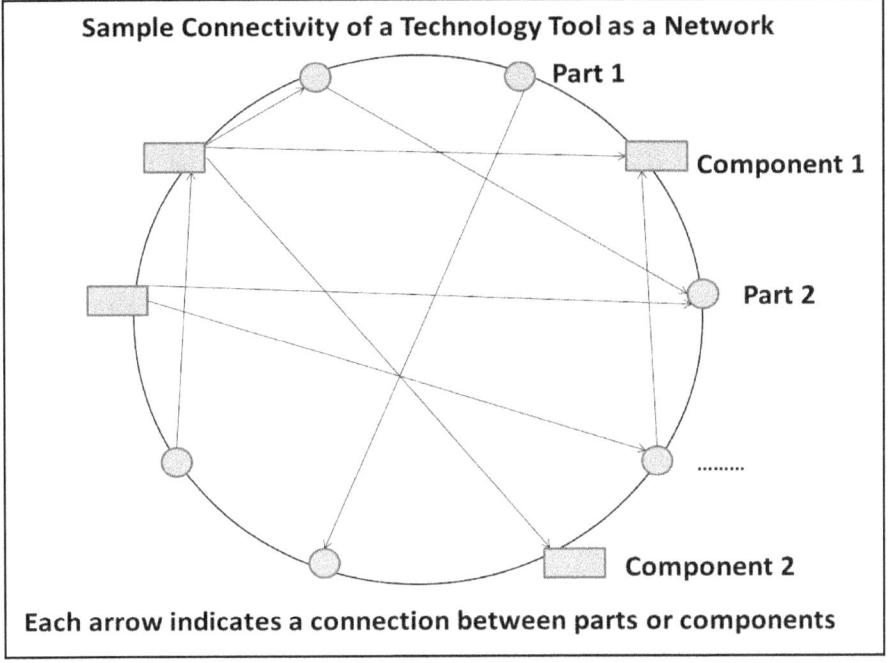

Figure 37 – Connectivity of a Technology Tool as a Network

Innovation in technology tools may alter these connections through modification of parts and components (revision, addition, deletion) as well as their individual connections. Such innovations result in improved delivery of an intended purpose or in some cases delivery of even altered purposes.

Similar to biological networks, some parts or components may have high number of connections to other parts or components in a technology tool. Any changes or disruptions in these parts or components may significantly disrupt the proper functioning of the technology tool. They increase the fragility or sensitivity of the technology tool in general.

On the other hand, the less or poorly connected parts or components may have only a small number of connections to other parts or components. Hence, they increase the robustness of the technology tool or a portion of it since changes to them may not cause significant disruptions or dysfunctioning in the technology tool in general. On the other hand, by and large they may not be highly critical for the functioning of the entire technology tool or even a portion of it.

The Shape (Morphology) of a Technology Tool

The entire shape and the overall body plan of a technology tool are determined by its designer. In some cases, they may be preserved by intellectual property rights and may not be replicated by other designers.

The entire shape and the body plan may be reverse engineered and identified in most cases and may even be replicated or modified if legally permissible and functionally feasible. Unlike the biological counterpart, there is no coding scheme embedded in the technology tool itself that depicts the body shape and the plan.

Also unlike the biological counterpart, replication or production of identical or similar technology tools require separate production (manufacturing) facilities.

Technology Tools - Purposeful Innovation

Technology Bang by definition occurred subsequent to the Life Bang. The design and production (manufacturing) of technology tools require other living beings. They are designed for an intended purpose and normally meet a well-defined need envisaged by their designers.

In reality, the technology tools may deliver the intended purpose for a period of time but then either the need may change (no longer needed) or the technology tool may be enhanced over time with novel superior designs, possibly with extended functionality achieving enhanced purposes. This creates a filtering out mechanism which replaces obsolete technology tools with new ones. The consumer[16] demand plays a central role in determining the adoption of technology tools and also the adoption timeframes.

If we look around the technology tools we use, we will see differences in terms of timeframes for using them. We might be living in the same house for more than a decade, using the same furniture and washing machine for a decade or so and they may still keep on delivering their intended purpose for us. On the other hand, we might replace our computers and mobile

---

[16] Consumer refers to technology tool consumer.

phones every few years due to extended functionalities and enhanced purposes delivered by them. These are just some illustrative examples and obviously there might be exceptions.

Incessant purposeful innovations create almost a recycling mechanism for technology tools. Those technology tools that do not adapt to changing consumer needs get eradicated while the successful ones that meet consumer demands continue their adoption.

Since consumer demand is external to the technology tool itself, it is imperative for the producers (manufacturers) of technology tools to strive for meeting the consumer demands and expectations. In a way, this is very much analogous for life forms (organisms) adapting to external environmental conditions.

Innovation Similarities and Differences between Life and Technology Bangs

It is quite fascinating to note that there are interesting similarities as well as particular differences between the products of Life and Technology Bangs.

Hierarchical Structure through Parts and Components: Organisms and technology tools are both made up of constituent parts and components. Despite the fact that origin of life[17] continues to be debated, it is commonly accepted that the cell as a fundamental component has appeared during the history of Earth which has the characteristics of life. Cells are fundamentally a chemical factory in which a significant number of chemical reactions, i.e. life processes, take place as a result of intra-cellular and extra-cellular stimuli and chemical materials. Cells are components (i.e. consists of a combination of several parts) which act as a platform for life and enabled diversification of different life forms (organisms). The cell has arisen at some point in the history of Earth at a certain intermediate level in the hierarchy of organisms.

---

[17] There are different theories which aim to explain the origin of life through terrestrial and extra-terrestrial means.

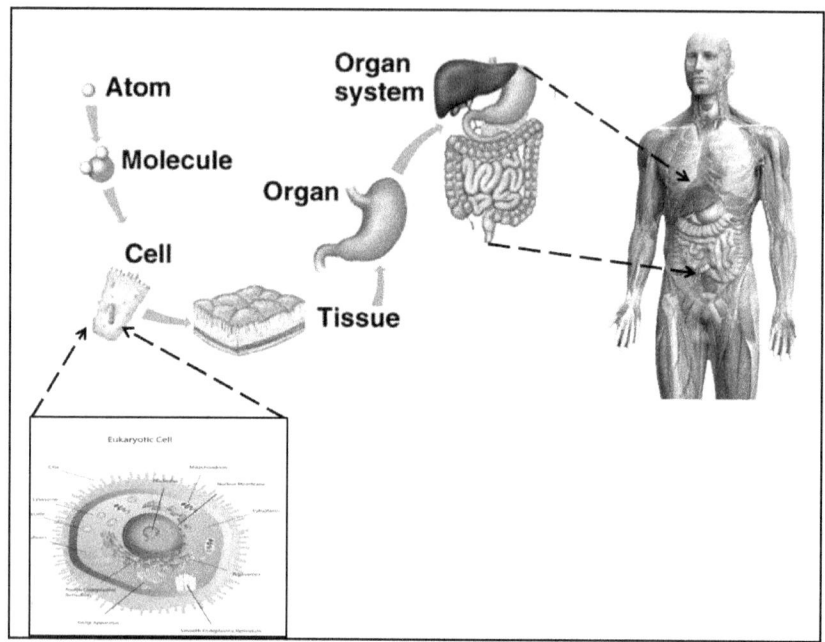

Figure 38 – Hierarchy of an Organism

Various life forms have the cell as their building block and are entirely made up of cells which differentiate based on the code in them.

Technology tools are similarly made up of parts and components from various domains such as mechanical, electrical, optical, etc. There might be multiple levels of hierarchy in technology tools assembling parts into components, components into further more complex components, etc.

Therefore hierarchical structure through a combination of constituent parts and components is common in both organisms and technology tools.

Functionality through Interactions between Parts and Components: We have seen in the Life Bang chapter that organisms form biological networks at multiple levels and these networks are connected at different levels with each other. The interactions in these biological networks determine the functions of an organism and can be quite complex.

Similarly, technology tool's parts and components interact with each other and their connections determine the functions and also the resulting intended purpose of the technology tool. Hence there is a strong similarity between organisms and technology tools.

The interactions between parts and components of organisms as well as technology tools may utilize transduction to maintain compatibility. Sensors and actuators can be used to ensure smooth communication and in general functionality between parts and components.

Nature of Interactions in Organisms and Technology Tools: Organisms are made up of cells which are essentially chemical factories that produce chemical materials and carry out various chemical reactions. The code in the cells (genes) produce proteins and the organism amasses various chemicals through nutrients and the external environment. The chemical reactions which collectively are called metabolism take place almost automatically which sustain the life processes in an organism. The important effect is that organisms are essentially chemical beings in nature and rely on chemical processes to maintain their existence in their environments.

Unlike organisms which are chemical in nature, technology tools are made up of parts and components which may belong to different domains such as mechanical, electrical, optical, etc. The design of technology tools in a way harnesses the laws of nature for different domains to deliver an intended purpose.

Design of Organisms and Technology Tools: Organisms are built from components called cells which act like a platform. The code which exists in all the cells clearly indicates how each cell will differentiate through the activation of certain genes in the cell. Furthermore, the code also indicates the body shape and the detailed design of an organism. Hence, the design of an organism and even its entire development from embryo stage to mature stage is all determined by the code in the cells of an organism.

On the other hand, the design of a technology tool is not embedded in the tool itself (nor in any portion of it). A designer (or an engineer) can reverse engineer a technology tool to determine its constituent parts and components along with their connections. Hence the design in general exists outside the tool itself (in general it is in the mind of the designers and may be stored as a drawing, digital file, etc.). In some cases, design of a technology tool can be kept as a trade secret and / or its intellectual property rights can be preserved through legal means.

Therefore design of organisms and technology tools differ considerably since an organism's design is embedded in each and every cell of the organism whereas a technology tool's design exists outside the tool itself and hence the tool itself is not capable of evolving itself over time but needs an external designer to make modifications on it. On the other hand, an organism has potentially the capability of evolving through modification of the code in its cells. This is a very important difference.

Reproduction of Organisms and Technology Tools: Organisms are capable of reproduction through sexual and asexual means depending on the specific organism. In fact, this is one of the characteristics of life. On the other hand technology tools need external production (manufacturing) facilities to produce them, so they cannot in general produce themselves on their own.

Innovations of Organisms and Technology Tools: Innovation in organisms and technology tools are quite similar to each other. The parts and components and the materials that they are made of as well as the connections (of parts and components) in organisms and technology tools are the sources of innovation.

In the case of organisms, the code in the cells (through alteration of genes) can inherently cause these innovations, or in some cases it may be triggered through external environmental changes. On the other hand, innovation in technology tools occurs as a result of design changes and is externally induced by the designers and consumers rather than inherent changes in the tools themselves.

In general, technology tool innovations are deliberate and they aim to deliver an intended purpose while meeting the needs and expectations of consumers. Conversely, innovations in organisms can happen through random alterations in the code within the cells, i.e. genes (the interested reader can further explore genetic innovations through mutations, duplications, recombination, transposition, sexual reproduction, etc.). As discussed earlier code changes in the cells of an organism (i.e. alterations in genes) can produce new chemical materials (proteins) and / or new connections in biological networks. These innovations are not deliberate and they do not aim to deliver an intended purpose; however the resulting organism needless to say will still endeavor to carry on its living state and possibly even gain an advantage through these changes in the context of external environmental conditions.

Hence organisms (i.e. Life Bang products) are purposeless innovations while technology tools (i.e. Technology Bang products) are purposeful innovations[18]. Therefore, purposeless and purposeful innovations co-exist on Earth leading to a staggering amount of material diversity. It is worthwhile to point out that Earth itself is a product of natural purposeless innovation which happened after Big Bang.

Some Futuristic Organism Innovations – Foresight through Insight

The insight gained in the previous chapters may allow us to do some projections for future innovations. In a nutshell, we have identified common as well as dissimilar aspects of Life and Technology Bangs related innovations. The parts and the components, their connections and the materials used in them are the main sources of innovations for both organisms and technology tools.

We have seen that the code in the cells contains instructions to control the production (manufacturing) of various chemicals which in turn partake in the life processes within the cells of an organism. They also contain instructions which regulate the development of an organism and determine its shape, body size and plan. Organisms including humans renew all their cells through division and recycle the old ones with the new ones periodically (unlike technology tools). For example our blood cells divide almost every 120 days (approximately three times a year) but our skin cells divide every 35 days (approximately once a month) and some other cells in our bodies divide every day. On the average we have about 100 trillion cells in our bodies, of which 500 billion are renewed every day. This so called cell cycle is important for our growth and sustaining our lives.

In general, organisms are mostly self-contained within themselves with the

---

[18] Scientific viewpoint has been taken in designating purposeful or purposeless innovations.

exception of some externally obtained chemicals (nutrients) and of course external environment interactions.

One of the differences on the other hand is that technology tool innovations are designed by humans (for practical purposes we will limit technology tools to human invented ones in this chapter) and therefore their knowledge resides in the humans or in the organizations that they established (companies, governments, etc.). The same cannot be stated for the organisms. They are not designed by human beings and their parts, components and connections have remained as a mystery for a long time and still continue to do so for many organisms.

However, we are on the verge of a major revolution in terms of organisms. For the first time in human history we are uncovering the intricate details of organisms. We are literally starting to look "Under the hood" in various organisms. It is as if we have just unearthed a spaceship buried deep in the desert the design of which we are completely oblivious. So we start to break it apart, try to understand its parts and components, which materials it is made of and how the parts and components are connected to each other in a very fundamental manner. And we reverse engineer it. This is very similar to our approach for understanding the organisms.

Needless to say, human curiosity in the past has analyzed organisms in terms of their anatomy and medicine has achieved significant successes. But the more recent change which is triggered by molecular biology and genetics is taking it into a higher and deeper level. Science is revealing the very parts, components and the connections inherent in organisms including the humans.

The Human Genome Project, which started in 1984, published its rough draft in 2000 and was declared complete in 2003. With that, the code in the cells of the humans was deciphered with a high accuracy rate and it turned out that human beings had approximately 25,000 genes which produced the chemical materials in the body called proteins. This gave us the parts that the human organism produced in its cells with the help of the code in the cells. This was a major achievement and it opened up new areas of research. Earlier, some scientists had believed that the complexity of human beings was related to the number of genes in the human genome but that did not turn out to be the case. Ironically humans have almost similar number of genes as the mice and the worms. It is now realized that the complexity of the humans arise from various connections in their biological networks we have seen in the Life Bang chapter. The metabolic, gene regulatory and the protein networks (among others) and their interactions within and among them make up the complexity of an organism. In other words, organisms may use similar number of parts and components, and even similar or the same parts in terms of structure and function. The complexity arises not only from the parts and components, but from all the intricate connections and interactions that take place among these parts and components.

Fixing Organisms Post-Ailment: In the past decade, tremendous amount of research has been conducted to decipher the code in various organisms, and to understand the biological networks associated with them. This is paving the way to highly important research projects related to inner workings of organisms. More importantly, it is uncovering the veil for some important diseases. It is, to a certain extent, offering new perspectives and fundamentally novel methods to treat malfunctions in the organisms.

More specifically, the root causes of some diseases are related directly to the code in the cells of organisms or to the biological networks associated with them. Cancer for example is a disease of uncontrolled division of cells. We have seen that every cell divides periodically in an orderly manner. In cancer, this orderly process fails and cell division occurs much faster than the normal periodic rate resulting in tumors. Biological networks and the interactions among different parts and components in them are shedding enormous light in the mechanisms of cell growth and division. The factors that influence cell growth and division are currently being identified and the mechanisms of various cancer types are being discovered. In some cases unintentional code changes in the cells (e.g. due to radiation) cause unintended chemicals (proteins) to be produced which malfunction and produce cancers. In some cases the connections among cells (through signaling or transduction) malfunction and produce cancers. Depending on connection specifics and the location in the body, different cancer types arise. It is easy to deduce that cancer is not just one disease but a group of several diseases depending on the mechanism and the location it occurs. So our deeper "Under the hood" understanding of cells and the organisms are opening new doors for therapy and treatment. We are starting to understand the root causes rather than symptoms only.

The technological advances have come to a point whereby we can intervene and make selective code changes within the cells. This can potentially fix the unintentional code changes in the cells and can avoid various diseases.

Similarly, malfunctions in connections among the cells may be repaired through prudent interventions. Needless to say, the biological networks are fairly complicated and our interventions may potentially affect other functions than the intended ones. That is why, a good understanding of the overall functioning of the biological networks and their parts and components are so critical to ensure targeted results.

In certain diseases, the organs and the tissues of organisms encounter irreversible damage. We have seen that tissues and organs are essentially combinations of differentiated cells. In the case of irreversible damage on them, the body naturally cannot fix itself. However, scientists and medical doctors are currently trying new techniques whereby undifferentiated cells (called stem cells) are being implanted in damaged tissues and organs. These undifferentiated cells start dividing and becoming differentiated based on the tissue and the organ that they are placed into. In a way, the tissue and the organ start recreating itself through the implanted differentiating cells. This is a natural solution for damaged body parts and it relies on the self-functioning nature of the cell. The undifferentiated cells start applying the instructions of the code in them and differentiate accordingly.

A fairly recent development has enabled researchers to form three dimensional brain tissues from stem cells in the lab. These tissues exhibit different regions of the brain and include signs of functional neurons. It is acclaimed as a major achievement given the complexity of the brain.

Similarly, another research team has been able to grow human heart tissue which beats by itself (autonomously) in a Petri dish[19]. It is quite interesting that the tissue itself was again formed from stem cells in a novel manner. The stem cells were derived from human skin cells which were reprogrammed to an embryonic state and then subsequently to heart cells. Even more interesting was that this new heart tissue made from human skin cells was then implanted in a mouse heart whose cells were stripped of

---

[19] A *Petri dish* is a shallow glass or plastic cylindrical lidded dish that biologists use to culture cells.

(essentially a heart with a scaffold) and after several weeks the mouse heart was fully rebuilt and it was beating on its own at a rate of 40 to 50 beats per minute.

Fixing Organisms Pre-Ailment: Understanding of parts, components, and biological networks can also assist us in developing preventive measures before an ailment occurs. Hypothetically, we can develop various kinds of sensors from different domains to gauge the activities in our bodies (currently we do some check-ups and selective tests on a periodic or needed basis). We can measure whether our life processes happen in the way that they are supposed to or not. Since we have the blueprint for the organisms (through the code, the parts and the components and the biological networks) we notionally know how normal organisms should function. We know what chemical products should be produced (e.g. protein production, metabolic reactions), how the cells should divide, how the code in the new cells should look like, etc. Knowing the ideal or normal values for those, we can devise ways to gauge all these activities closely and determine any discrepancies at early stages before they escalate to an alarming level which we call disease.

New technology tools can be developed using existing and novel parts from mechanical, electrical, optical, nanotechnology domains to do these tasks. To illustrate the concept with a far-fetched example, think about a nano-size (extremely small, much smaller than the width of your veins and cells) robot that wanders in your body and has delicate sensors to measure various chemical, temperature, pressure, acidity, etc. levels in various parts of your body and sends them electronically to a device (a sophisticated computer). We can see the state of our bodily functions and identify any abnormal conditions detected by the sensors almost real-time (right at the same time). This may allow us to take precautionary measures almost instantaneously to avoid further impact on our bodies. Nano-size robot was given as an example; in some cases we can use remote sensors without necessarily injecting foreign devices to our bodies. These might have seemed like science fiction a few decades ago but today science is closing the gap at an accelerated pace and they might become a commercially available reality since there are several trials in research labs.

Cross-domain Innovation in Organisms: Organisms are essentially chemical in nature and use the cell as a platform to build upon and rely on biological networks for their functioning. There is a chance that in the future some of the functionality in the organisms can be provided or enhanced by mechanical, electrical, optical, and nano-technological parts and components among others. Organisms can incorporate the best of all domains in their functioning. Potentially new sensors, actuators, diagnostic and control components can be integrated in organisms. The opportunities are countless in this area since organisms have so far been exclusively chemical in nature. But the potential is there to augment it with parts and components from other domains.

An Inventory of Organisms: At the moment, we are discovering the inner workings of organisms to a never before seen level of detail. We are enthralled by the variety of codes in organisms, the parts and components utilized and the biological networks associated with them. It is in fact quite interesting because we can analyze not only the living organisms but the extinct ones as well. The fossils of extinct organisms enable us to decipher the code in their cells, understand the chemicals (proteins) produced in their cells, and even make projections about their biological networks. This can potentially create a detailed archive of all the organisms that existed after the Life Bang (at least the ones whose relics are still available) and record their progression over time. This may give us a good idea about the similar as well as distinctive features of organisms, what pieces of code, chemicals and biological networks were preserved among organisms. It might also possibly give us an idea about the environmental conditions that they were in.

Smithsonian Institute has already embarked on an ambitious project called Global Genome Initiative to preserve and study the genomic diversity of all the organisms and provide the genetic information to researchers worldwide. This will significantly contribute to understanding of extinct and living organisms on Earth which are the products of Life Bang.

Creating Organisms[20]: So far, we talked about analyzing organisms to understand their inner workings, and to use that knowledge to cure or

---

[20] This section is a highly controversial section and the ideas in it may raise significant ethical concerns; they are covered to illustrate the possibilities that arise with innovations in organisms.

prevent ailments, and to enhance them through cross-domain innovations.

As we all know, some organisms are extinct. They have existed sometime during the history of Earth. For some of those organisms, we have access to their codes in their cells through their fossils. This opens up an interesting theoretical possibility to literally create them again. The new field of biology, called synthetic biology goes beyond mere analysis and allows synthesizing a new organism by designing its code and implanting it to a cell by replacing its original code with the designed code.

Organisms have emerged over time as opposed to technology tools which were designed by humans. Synthetic biology changes this undisputed fact for organisms. For the first time, we are in a position to synthesize new organisms. We are accustomed to doing this for technology tools for millennia and our craftsmanship for technology tools has advanced substantially. Conversely, we are at an infancy stage for synthesizing organisms. The first success was declared in 2010 whereby a bacterium code was synthesized outside the cell (genome of a new organism) in a digital computer and was inserted successfully into another cell and replaced its code. A new bacterium organism was created through its code design. This was the first synthetic life and was successful. For some, this may be considered playing God and for others a negligible microscopic achievement. It was accomplished after more than 10 years of massive efforts by a team of 20 scientists. The result was a microscopic organism which cannot even be seen by our naked eyes, however the implications were extraordinary. A novel organism had for the first time been created by another organism from scratch through synthetic design. Until then, organisms had created novel technology tools but not novel organisms. So this event marked the beginning of a new era in which new organisms can now be synthesized in addition to their natural course of progression.

There is a chance that synthetic biology may start creating new organisms with various intended purposes. The code, the cell (as a platform component) and various chemicals can be combined to produce (manufacture) new organisms to deliver certain intended purposes. Needless to say, our understanding is currently shaping for various biological networks. So we are more likely in an analysis or understanding phase than the synthesis or design one. But at some point in the foreseeable future, we will grasp the fundamentals and the intricate details of metabolism, gene regulation, and protein interactions. We will map them to their constituent parts and components and then enhance our synthesis or design skills for organisms. Just like we did for technology tools, we will most probably develop craftsmanship for organism designs. This is definitely a new research and application area with promising potential. In fact, the International Genetically Engineered Machine (iGEM) Foundation[21] is a good example of such efforts. iGEM organizes competitions every year among international high-school and college students to design new genetically engineered organisms by using a registry of standard biological parts[22]. As this clearly demonstrates, biological parts are already being treated as technology parts and components to be used in designing purposeful organisms. The projects submitted in iGEM competitions are already mind-boggling examples and are a testimony to our fast pace in developing craftsmanship in novel organism designs; especially considering that they are designed merely by students.

The remarkable and promising aspect is that the code design provides us an ample opportunity for innovation. The code in the cells produce chemical materials (called proteins). We have seen that humans for example have approximately 25,000 protein producing pieces of code in their cells. A protein is also an innovation through combination whereby different chemicals called amino acids are combined together (the way these amino acids combine are subject to the electromagnetic fundamental force we have seen earlier). Each amino acid is like a bead in a necklace, the necklace representing the protein. There are more than 20 different bead types (amino acids) and each necklace (protein) may consist up to hundreds of beads (amino acids). This creates an unimaginably huge number of potential necklaces (proteins) by combining different types of beads (amino acids), by changing the order of the beads and by changing the total number of beads. The organisms that we see on Earth have barely scratched the surface of this vast potential. We are confronted with this potential right now and there is so much to explore and to discern as to which of these huge potential number of proteins would create novel functionality in organisms. Once we determine these new proteins and their functionalities, we can

---

[21] More information can be obtained from www.igem.org.
[22] More information can be obtained from parts.igem.org.

then create their codes in the cells to synthesize them.

We have just stepped into a new era in which we can design the code for cells in our computers (predominantly an electronic domain technology tool) and the code will have all the instructions needed for an organism; i.e. it acts as a piece of software for creating a new organism. Then we can insert the code in a cell. The cell acts as a hardware platform to run our software on it, i.e. the code. Hence, a new organism can be created from scratch. Of course, the entire process and its detailed steps are oversimplified in this explanation but the basics are quite similar.

A few months ago, I downloaded a piece of software from Internet called a compiler. From my early days of exposure to information technology, I was quite familiar with compilers and their designs (in fact my undergraduate capstone project was to design an assembler which was a rudimentary version of a compiler). But the compiler that I downloaded from the Internet was a very different one. It was called Genome Compiler. It helps in designing the code for cells, i.e. designing and debugging a genome for an organism. It looks a bit like computer aided design software applications, not for mechanical technology tools, but for organisms. It allows you to design your own code for an organism. Genome Compiler is an example of a novel piece of software (technology tool) which can be used in synthesizing new organisms.

One of the first practical applications of Genome Compiler was to use it to design a glowing plant (through green fluorescence) which was a first step in sustainable natural lighting according to the owners of the idea[23]. It relies on changing the code of a plant (modifying its genome) to synthesize a green fluorescent protein in the cells of the plant and in turn produce natural light. The entire idea has caused quite a bit of controversy due to the fact that it entails production of a novel non-natural synthetic plant as a new organism.

In the world of technology tools, 3D printing seems to be a potentially disruptive innovation that may utterly alter the way materials and end products are produced. It may democratize production of technology tools

---

[23] Glowing plant has been posted as an idea in kickstarter.com and has surpassed its funding goal in 2013.

by allowing them to be printed at home through 3D printers. It seems likely that we might in the foreseeable future start printing 3D materials as opposed to papers only. But concomitantly, there is another development taking place in the biology field. New 3D bioprinters are being designed. 3D bioprinters print tissues and organs. They are essentially biological printers and produce biological parts and components. The bioprinted tissues and organs can be used in surgery, transplantation and pharmaceutical applications. They are examples of synthesized biological parts and components that make up an organism.

Another biological printing type currently under development is DNA printing. DNA printing prints the actual chemical material including the storage of the designed code to be placed in the cells (referred to as DNA). Currently, synthesizing the actual biological code in the cells is excessively expensive. This is very much similar to initial printing costs in seventies and eighties for computer printers which were highly expensive. It is projected that the DNA printing costs will also come down significantly over time just like the computer printer costs have in the past. DNA printing may then become affordable and it will make creating novel organisms much more feasible financially. Today's code synthesis technology in 2013 incurs a cost in the order of billion dollars (or hundreds of millions of dollars) to print a designed human code and may take really a long time.

All these advances are the bits and pieces that collectively enable our progress towards creating organisms. We are currently far from mastery of craftsmanship skills for modifying or creating organisms. However recent developments are a testimony to our slow but sure progress in that direction.

Demarcation between Organisms and Technology Tools

The futuristic innovation examples discussed in this chapter rely on the same fundamental innovation types and mechanisms which we discussed in the previous chapters. The cell as an important component (life's building block and platform) and the code in it play a very important role in these innovations.

In this book, we have separately discussed the organisms and the technology tools, partly because there has always been a clear demarcation between the two. But we have also seen that there are fundamental similarities between the two in terms of innovation. In fact, new advances such as synthetic biology are starting to blur this well-established demarcation between technology tools and organisms. This may sound as a strong statement because we are only at the beginning of this new era and the demarcation between natural organisms and man-made technology tools is still very much pervasive at the time of writing this book. However, we can agglomerate both the technology tools and organisms into a single logical grouping in terms of innovation due to their similarities.

This allows us to end up with a new enriched and enhanced toolbox for innovation composed of more domains; namely chemical, mechanical, electrical, optical, etc. This new toolbox gives us more parts, components, materials, and connectivity options. Novel combinations can be discovered producing new functionalities and intended purposes. Depending on particular design constraints and performance characteristics, we can employ the appropriate domain or domains in our designs. We can mix and match them as needed, convert between them through transducers.

We can question the viability of our existing technology tools, their parts and components and possibly replace them with organisms. Future homes and furniture might be synthetically designed organisms with different shapes. If this trend picks up it would be quite difficult to distinguish technology tools from organisms in the distant future and treat them separately as we did in this book.

Alternatively, we can utilize electrical (electronics) domain tools to control technology tools and organisms. Certain intelligence can be built in most technology tools that we use and they can be interconnected to each other for communication purposes. It seemed always intriguing to me why the furniture that we use in our homes is so inflexible and has a single spatial configuration. Such examples can be enumerated and elaborated (e.g. Internet of things).

We have earlier seen that platforms as special components have a catalyzing effect on innovation. They tend to increase the innovation potential several fold. The cell as a building block is a special platform that has built all the organisms (life forms) on Earth. One wonders if the cell types that we know are the only viable ones. The cell in the hierarchy of innovation sits at a certain complexity level. It is definitely not the most fundamental part or component compared to a single chemical element or compound. It possesses several chemical materials (e.g. water, ions, proteins, nucleic acids, phospholipids, etc.) and is bound by a membrane. So it is just one component out of many possible ones. But all organisms as a platform have utilized it.

Perhaps, we can now build an enhanced version of the cell as we know it, i.e. another more advanced chemical factory with diverse and distinctive functions in it (possibly containing parts from other domains as well). We can build new codes for it producing different chemical materials with different biological networks. We can even potentially change the way the code is stored in the new enhanced cell chemically (i.e. a new chemical material replacing the DNA in the cells). This new enhanced cell can become a new platform for completely different organisms. In other words, we use the same fundamental rules of innovation to create new components and platforms. We can combine the above mentioned cell related enhancements to produce diverse cell types with varying complexity levels leading to multiple cell platforms as opposed to existing ones. Each cell type can be a spring board for new life forms (organisms). This might almost qualify as a potential candidate for a new Bang, an additional one after the Technology Bang, or might be considered as an extension of the existing Life Bang.

Platforms can on the other hand potentially curb and even restrict innovations. The very ease which enables further innovation on top of an existing platform may reduce other innovations to replace or to alter the platform itself. Organisms have used the cell as the building block of life and alternatives to existing cell types have not transpired throughout the history of Earth. We see similar trends but with different timeframes in processors, operating systems, etc. A platform tends to divert the progression of related technology tools and organisms deeper into a certain path which may stifle innovation in the long run due to excessive dependence on the platform itself. But platforms due to their ease of use and convenience have triggered proliferation of related technology tools and organisms. Hence, real life evidence strongly supports the fact that innovations tend to capitalize on platforms.

The new enriched toolbox amplifies our innovation potential and provides us a new outlook. Some of the futuristic innovations we discussed blur the demarcation line between technology tools and organisms. We have a vast potential of combining different parts and components from both sides of the line in myriad ways to deliver novel intended purposes. We currently have a better grasp of technology tools since we have designed them. Our knowledge about organisms is still being established and relatively more limited at the moment. As we excel in our understanding of organisms, it is likely that the demarcation line may gradually disappear. We may treat organisms and technology tools as mere contenders in our enhanced toolbox for innovation. The opportunities as well as the challenges are enormous in this new wave of innovation.

A Short Treatise on Other Innovations

In this book, we have discussed how various visible matter in the Universe was formed as a result of three different Bangs. Combination of various constituents (parts and components) was at the root of all the innovations. Celestial objects, organisms and technology tools were all formed from various parts and components. These parts and components were interconnected physically and in some cases through the fundamental forces of nature. The utility of the innovations regarding visible matter rely on their tangible aspects.

Intangible Innovations: It is quite interesting that there are other innovation types the utility of which rely on intangible aspects such as information. The information present in these innovations can be a combination of text, images, shapes, colors, odors, feel, sound, etc. which are aspects that pertain to our senses. An example might be a book whose utility relies mostly on the information in it. This information can be a combination of words, sentences and possibly images. Similarly, various art forms such as paintings, sculptures, performing arts, etc. are all combinations of colors, materials, shapes, sound, images and others which convey information that evoke certain emotions, feelings and have aesthetic values. The intangible aspect is the primary one in addition to the tangible features present in them (the physical book itself, the canvas and paints on it, the bulk of the sculpture, etc.). In addition to various art forms, we can also treat alphabets, languages, symbols, mathematics, laws, regulations as intangible innovations which can be decomposed into their constituent parts and components. Hence a category of intangible innovations exists which can also be analyzed and modeled as combinations of various parts and components. We will not study and scrutinize intangible innovations further in this book since our main focus has been on the visible matter and its related innovations. But it is quite noteworthy that the same basic principle of combination applies to intangible innovations as well.

Enabling Innovation: On Earth, organisms have diversified due to natural purposeless innovation which mostly takes place through code changes in the cells. On the other hand, technology tools are purposeful innovations and are conceived by humans.

The production or manufacturing of technology tools also require a combination of parts and components just like the technology tools themselves. These parts and components can be organisms (humans for example) or other technology tools. Humans have for example created organizations to produce technology tools. These organizations can be companies, government entities, NGOs (non-governmental organizations), etc. The organization itself is an abstract entity (intangible) and it is composed of people (organisms) and various assets (e.g. buildings, machines, furniture, other tangible assets, etc.). The assets are technology tools themselves. Hence an organization can also be modeled as a combination of organisms and technology tools combined through a special configuration. Unlike technology tools whose constituent parts and components are inter-connected physically or in some cases through the fundamental forces of nature, the organisms and technology tools in organizations may have soft connections, i.e. interactions. These connections are not necessarily hard wired or established through the fundamental forces of nature. They are impermanent connections and may involve temporary interactions among the parts and components. For example, humans in an organization may interact together through discussions, meetings, etc. to design other technology tools and may then utilize machines to manufacture the designed technology tools. All these are temporary interactions. Humans may leave or join organizations; new technology tools can be acquired or similarly disposed in an organization. But at a given point in time, we can model organizations that produce technology tools as composite entities combining organisms and technology tools to deliver an intended purpose. The inter-connections in the organizations may be impermanent and soft through various interactions as opposed to relatively long-lasting and hard (physical) connections in technology tools. Hence an organization can be modeled as a combination of parts and components as well whereby the parts and components can themselves be organisms or technology tools.

The structure of economies can further be modeled as a combination of organizations such as government entities, private sector entities and others.

So we see a fairly hierarchical network like structure in economies composed of various constituent organizations. The interactions among organizations make up the activities in economies. The interactions among organizations can be conducted through human to human or in some cases through electronic interactions. Internet for example as a technology tool has enabled various interactions among organizations to be conducted electronically (e.g. obtaining information, purchasing products and services, providing customer support, invoicing and payments, etc.) in addition to human to human ones.

In a way, one can think of organizations as platforms in economies. It is a very commonly used component and has progressed significantly over time. Specialization in economies has enabled proliferation of various private and public sector organizations as highly used components which focus on delivery of specific products and services. Organizations as platforms rely on other innovations hierarchically. As an example, they rely on other technology tools (tangible innovations) which are used as assets within the organizations; they also rely on commercial laws which are intangible innovations regulating various activities and determining rules of engagement among organizations.

Organizations play a major role in economies both as a platform and therefore naturally as a major constituent. Similarly, the society can be modeled as a combination of individuals (organisms) and related constructs such as families, groups of individuals, etc. in a hierarchical manner. The interactions among individuals and social constructs make up the activities in societies. The interactions among individuals can be conducted through human to human or in some cases through electronic interactions. Internet as a technology tool has enabled various interactions among individuals to be conducted electronically (e.g. social media) in addition to human to human ones. The interactions in society tend to also occur through soft and impermanent connections. Formally established social groups and organizations can be thought of as platforms in societies just like organizations in economies. Hence, economies and societies have analogous hierarchical structure and fundamentally analogous interaction mechanisms.

It is easy to discern that economic and social innovations are both purposeful innovations. Their interactions are deliberate. Organizations are established to serve an intended purpose and they interact among each other for well-defined purposes. Similarly, individuals and social groups interact intentionally and to achieve certain intended purposes.

Combinations in Innovations: We have started our journey in this book by stating that "Innovation through combination subject to natural laws" will be the common unifying theme of this book. We have also seen that micro to macro scale innovations are combinations of parts and components. Protons, neutrons, atoms, molecules, compounds, matter, stars, planets, galaxies, cells, tissues, organs, organisms, technology tools are all combinations of various parts and components.

The combinations in these innovations have different characteristics. Quarks are combined to form protons and neutrons and this combination happens through the strong force which is one of the four fundamental forces of nature. Atoms are combinations of electrons, protons and neutrons and this combination happens through the electromagnetic force which is one of the four fundamental forces of nature. Similarly molecules are combinations of atoms and compounds are combinations of molecules and they all combine through the electromagnetic force. On the other hand large scale celestial objects such as stars, planets and galaxies are combinations of various molecules and compounds and this combination happens through the gravitational force which is also one of the four fundamental forces of nature. Similarly, in organisms, most interactions are chemical and physical ones which boil down to electromagnetic force. In all these examples combinations are handled through the fundamental forces of nature.

In tangible technology tools, combinations of various parts and components are through physical hard-wired connections. Various mechanical, electrical, optical, etc. parts and components are connected physically together at the macro level (unlike through forces). If we look around us, we will see this clearly in almost all technology tools.

In intangible innovations such as pieces of art, language, laws, etc. information is established through a combination of text, images, shapes, colors, odors, feel, sound, etc. which are aspects that pertain to our senses. In all these, combinations may reside as information on a technology tool such as a novel, scores of music for a symphony, colors and images in a painting, a book of laws but the whole (aggregate) is the real innovation and carries some information which is perceived through cognition processes in the brain. Hence, the combinations are logical as opposed to physical.

Organizations in economies and social groups in societies on the other hand are combinations of organisms (e.g. human beings) and technology tools as assets. The combinations happen through various interactions. These interactions are human to human, human to machine or machine to machine in general. The discussions, meetings, decisions, etc. entail human to human interactions in organizations and social groups. They tend to be logical ones relying on information exchange. Depending on machine interface, humans and machines also interact with machines and provide physical stimulus or information to operate them (technology tools such as computers, robots, etc.). Hence interactions are the combinations in these innovations.

Therefore we see a wide range of combination types available for innovation. Different innovation types have utilized different combination types.

Designing Innovations: Regardless of scale, from micro to macro, combination of parts and components has enabled innovations with or without purpose. The same will persist in the foreseeable future. Having partially but reasonably mastered technology design craftsmanship and having started to acquire organisms design craftsmanship, human beings are in a position to innovate further. Our toolbox for innovation is exceedingly becoming richer with novel additions in terms of potential parts and components. This further magnifies the number of potential combinations in our toolbox. The innovation challenge becomes one where a purpose (including all design specifications) is met through a unique combination of parts and components in our toolbox.

It is fascinating to consider that any innovation is in fact a new combination (possibly a new configuration as well) of existing parts and components in our toolbox. From a conservative perspective, one can allegedly claim that there is nothing new in the universe since we are just reusing (in a way recycling) the existing matter in the form of parts and components and repurposing them. But from a progressive perspective, we are creating novel technology tools with novel purposes. It probably depends on the viewpoint. The late historical trend (last few millennia) indicates that innovations have proliferated and enhanced thanks to progresses in cumulative knowledge and craftsmanship. Especially humans have capitalized on existing knowledge and have found myriad ways to combine the existing innovations to further design new ones. It is not hard to predict that the fast pace of innovation will continue, possibly even in an accelerated manner in the foreseeable future as well.

Concurrent Innovation at Different Levels

At the beginning of the book, the main unifying theme was stated as "innovation through combination subject to natural laws". We have seen that innovation occurs at different scales in the universe.

At the macro scale, or universal scale, we have seen that galaxies composed of stellar systems form the main hierarchies throughout the universe. Planets in stellar systems are the major candidates for relatively benign conditions for the emergence of life. So they form the oases for innovation. Stars are the furnaces and sources of energy for innovation on the planets. Stars from their birth to date present a window of opportunity to their planets for the emergence and progression of life. Depending on star, this window of opportunity may last from tens of millions of years to several billion years. Needless to say, the longer the window of opportunity, the higher the chance of emergence and the progression of life is. Universe at a large scale keeps on recycling its large scale structures; i.e. stars and galaxies along with their planets. Furthermore the death of certain stars provides a window of opportunity for manufacturing heavier elements and ejects them into space which could be used as raw materials on other planets. The heavier elements produced can enrich the toolbox used for innovation on other planets. From a macro scale innovation perspective, the birth of stars concomitantly creates an opportunity for its planets to form and evolve life on them which in turn would fuel innovation. On the other hand, the death of stars also discontinues the life on their planets bringing it to an end, but produces heavier elements which will subsequently support life somewhere else in the universe. Hence macro scale innovation creates temporary opportunities for the emergence of life and innovation all over the universe.

If we think about the universe and the various planets in stellar systems and galaxies where life and technology can potentially thrive, innovation might be happening simultaneously all around the universe (even though we have yet to discover its evidence). We can think of it as an astounding co-experimentation at large scale in the universe. The planets on which life and technology thrive can be thought of as isolated islands where this innovation co-experimentation takes place. Earth has so far been isolated since we are unaware of the existence of life supporting planets and we have had no contacts with them. There is a chance that some other life supporting planets might be in contact with each other or perhaps Earth is the only life supporting planet in the entire universe (despite the probability being low at least statistically due to the sheer number of galaxies and stellar systems in the Universe). We do not have the answers yet and we have been so far on our own in the entire Universe.

We have seen in the first chapter that the large scale innovation follows the laws of nature and most large scale structure formations happen due to gravitational forces in the Universe. Hence, we have identified it as natural purposeless innovation. The laws of nature, basically the four fundamental forces, play out their course and the material in the Universe follows suit. We have a good understanding of the mechanisms involved in macro scale innovation and our understanding has been supported and verified so far with our astronomical observations.

The Big Bang and the laws of nature have created our dynamic universe. At a macro scale, the matter coalesces at various locales in the universe temporarily and forms the large scale structures in our dynamic universe. Hence, we observe "innovation through combination subject to natural laws" at the macro scale.

The planets, as the oases or the bedrocks of innovation, play a special role by enabling further innovation through emergence of life, which in this book we referred to as Life Bang[24]. Innovation through life rides on the macro scale innovation and needs it as a prerequisite. So far, we have evidence of life only for one planet, which is our Earth. Our understanding of life is based on this single occurrence and may most likely be insufficient to grasp different life forms in other planets (assuming they exist). As we have seen in Chapter 2, Earth is as dynamic as the universe itself and has gone through major changes during its history. Significant changes have

---

[24] We have oversimplified things by making planets as the only celestial bodies suitable for life. In reality, even satellites of planets, like the Moon, are potential candidates for life. Even the asteroids or other smaller celestial objects can potentially accommodate life.

taken place on Earth's crust and in its atmosphere. Life, as we know on Earth, has adapted to changes on Earth, in some cases went through partial extinctions but has been able to survive over billions of years.

The origin of life has not yet been identified scientifically. There are several competing hypotheses and theories trying to explain it. It is safe to assume that at some point during Earth's history, cell has emerged as a complex chemical component, rather as a platform, and as the building block of life. Therefore the first organism had emerged and from then on life has flourished for billions of years. The cell as a platform has supported innovation in organisms and millions of different organisms have transpired on Earth. The staggering diversity of organisms on Earth is all based on the cell as a platform and, the code changes along with different connections in biological networks have been the main source of innovation in organisms. So the Life Bang has unleashed the second innovation at the planet level in addition to the macro scale innovation at the universe level.

We have seen that primarily carbon but also several other chemical elements have played a crucial role in forming the chemical materials that support life processes. For example, the special chemical material called DNA has been the only source of code in the cells of all living organisms on Earth. DNA is the main source of replication and inheritance for organisms. We still do not know if DNA based life is the only life form in the universe or not. Similarly, the cell has been the main platform and the only building block for all the diverse organisms on Earth. Again, we do not know if all life forms in the universe would use a similar cell platform or not. Potentially, there might be other life forms in the universe which are neither DNA nor cell based. We have more questions at this point about life than we have answers. This is a mere result of our lack of knowledge about other life forms assuming they exist in the universe.

Life has been able to utilize some of the heavier elements produced in deep stars during their deaths in addition to the lighter elements produced right after the Big Bang. Therefore the chemical materials that life utilizes (e.g. sugars, proteins, nucleic acids, water, etc.) are all combinations of simpler chemical elements that have been produced earlier in the universe and coalesced on Earth. Unlike the macro scale innovation which is mostly based on gravitational force, life has capitalized on the electromagnetic force immensely to carry out its chemical reactions and interactions leading to an enormous variety of chemical materials. The cell as a platform for innovation combines some of these chemical materials in a certain way and uses the code in it to regulate the production of other chemical materials. Similarly, the biological networks that we have seen are combinations of all these chemical materials and their interactions. So once again life as a major innovation mechanism relies on combination of various parts and components while complying with the laws of nature. In general those organisms that have adapted to and survived in their external environments have been successful and have sustained their existence. This survival is not a purpose per se but rather a consequence of their capability to fit into their environments and adapt to the external environment conditions. Hence, life in general also relies on natural purposeless innovation since no explicit purpose has been scientifically identified through concrete evidence.

The third and the last innovation that we have seen is the result of Technology Bang and relies on both macro scale and life innovations. In other words, the Big Bang and the Life Bang are prerequisites for it. It encompasses the technology tools designed by organisms (for practical purposes we will discuss the human designed ones). Technology tools as we have seen are composed of various parts and components and in general deliver an intended purpose. So once again, technology as a major innovation mechanism relies on combination of parts and components subject to natural laws. The parts and components in technology tools can belong to various domains such as mechanical, electrical, optical, nanotechnology, etc. Since technology tools are intrinsically designed for a purpose, they are considered to be purposeful innovation. They are not produced naturally through the application of laws of nature. But they are rather produced intentionally by explicitly and consciously utilizing the laws of nature and they deliver an intended purpose. They need an organism as a designer (in most cases human beings but there are examples of other animals that design technology tools).

Chronologically, Life Bang has followed the Big Bang and the Technology Bang has followed the Life Bang. All three Bangs are required to explain the material diversity that we see in the universe. Each Bang has contributed in a special way to the material diversity that we see around us.

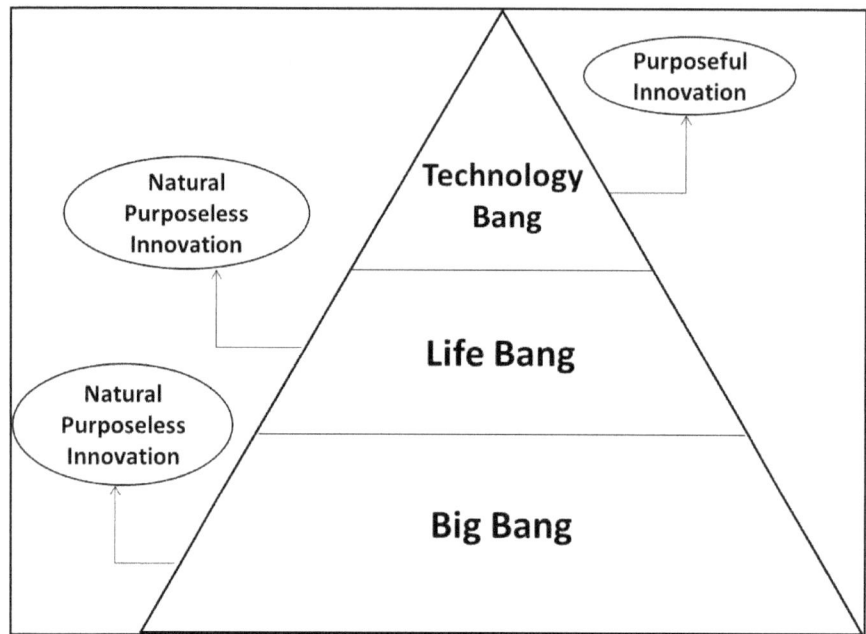

Figure 39 – The Cumulative Bangs

Organisms that emerged as a result of Life Bang have produced the technology tools after the Technology Bang. The interaction between organisms and technology tools was always unidirectional. Organisms designed technology tools for an intended purpose, but not vice versa in the past. Furthermore, organisms did not create other novel organism types. Organisms were created by natural means without the intervention of other organisms. These statements have been accurate for billions of years on Earth up until recently.

But not long ago we have entered into a new innovation era since we are equipped with new technology tools and understanding regarding organisms and life in general. Now for the first time in the history of Earth, humans as an organism type are in a position to utilize various technology tools to modify other organisms including themselves and even to design or synthesize novel ones. This is an uncharted territory and is being shaped as this book is written. The natural purposeless innovation mechanism of life which has hitherto been valid is currently being invalidated. We humans are now becoming capable of designing purposeful organisms through the usage of technology tools.

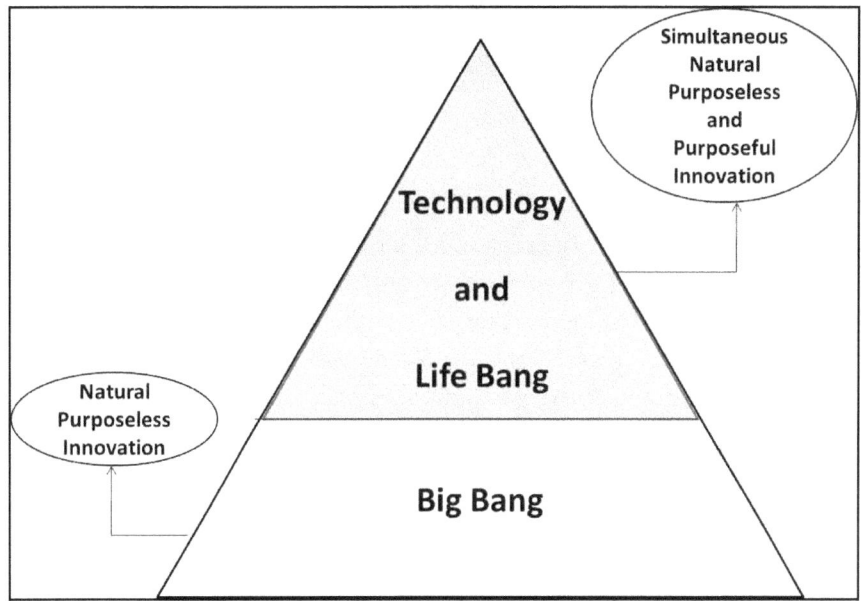

Figure 40 – Convergence of Technology Tools and Organisms

Moreover, the clear cut demarcation between organisms and technology tools are disappearing since we are in a position to interchangeably use technology tools as portions of an organism or vice versa. In other words, our innovation toolbox, which is composed of parts and components, has been greatly enhanced with the inclusion of technology tools and also organisms (whereas previously it was almost entirely composed of technology tools). We can now create organisms and technology tools to deliver an intended purpose. Needless to say, we are at the infancy stage of this convergence. However, several enabling bits and pieces are either being availed or under construction. This is really an exciting but also an intriguing development. We are confronted with an incredible potential that has implications to radically alter our existing technology tools and even the organisms. Yet it also poses great challenges including ethical ones. Time will tell how we will make use of this new potential. But it can be qualified almost as a milestone on Earth due to its lack of presence hitherto.

We are now in a position to decipher the inner workings of organisms since we are really going "Under the hood" and discovering all the intricate details. This is a bit like reverse engineering what was not manufactured by us. As we unveil the mechanisms and the parts, we also discover how we can modify or even design novel ones. Our craftsmanship in technology tools may not directly apply to organisms. Given enough time, we will also master those skills and even combine the two together to carry our innovations to a higher level.

I still remember the early days of information technology (IT) including personal computers and Internet and the potentials they posed. They were precarious yet exciting. I vividly recall using the first pre-release versions of Web browsers and experimenting with them before they were even officially released. Staying awake for a few nights, my curiosity led me to write simple programs to stream audio files over a local area network from one computer to another back in early nineties. The potentials that were envisaged then have all materialized over the years and now have diffused to our everyday lives in our homes and workplaces. The limitations of early days have all been overcome and dream technologies and devices of late eighties and early nineties have become commonplace today.

The potential of the current new innovation era reminds me of the early days of computers and Internet. There are daunting challenges, yet there is optimism that they can all be overcome. The benefits from this new innovation era are incalculable. It may trigger a revolution in the way we design our current technology tools. We may adopt bio-design techniques for various parts and components we use in our technology tools. And we can design new organisms for novel purposes.

When we look back at the last few centuries we realize how much we have progressed from industrial revolution till today. The pace of innovation has accelerated significantly. Given that we still have a few billion years for organisms on Earth before the Moon or the Sun (or potentially another celestial object) causes extinction events (barring other unexpected extinctions), it is safe to assume that innovation will continue and will reshape both the organisms and the technology tools on Earth. Maybe your guess is as good as mine. I do not want to speculate what might happen, but I have a strong feeling that we are about to experience another major transformation. We are living in exciting times.

The three sequential Bangs which unfolded over the course of Earth's history is almost rolling back now. The technology tools created as a result of Technology Bang is now starting to change the organisms which Life

Bang has created. The unidirectional and cumulative development among the Bangs is changing course once and for all. The technology tools are now being utilized to reshape and redesign organisms in a novel manner. As the clear demarcation between Life and Technology Bangs gradually become more ambiguous than ever, novel and distinct material forms are likely to emerge.

We can think of this era as one where the organisms and technology tools are almost inter-mingling rather than remaining distinct. The unidirectional cumulative build-up is not true anymore. Technology tools have almost reversed direction and started reshaping the very organisms which created them. We are bound to experience novel innovations which we are not accustomed to.

In this book, we have treated planets as the oases or bedrocks of innovation. We have taken macro-scale innovations (e.g. galaxies, stellar systems and the planets) as natural purposeless innovations which were the manifestations of natural laws in the universe.

However, humans have started using technology tools to alter organisms on Earth. In the future, we might gain the possibility of causing Life Bangs on other planets (or more generally on other appropriate scale celestial objects such as other planets or satellites of planets). Today, we lack such advanced technologies but it is probably a matter of time in the future to develop them and initiate a Life Bang which may in turn instigate a Technology Bang consequently. This is a bit like replicating what we have experienced on Earth in a different locale in the universe. The external environment conditions on such locales may affect organisms and may potentially produce distinct and uncommon organisms compared to Earth bound ones. Hence, we would partially shape the concurrent innovation that takes place at the universe level.

One of the current restrictions with our technology is that it has been limited to developing technology tools and has very recently started changing the organisms. But it has not been sufficiently powerful and advanced to change the macro-scale innovations. In other words, so far we have not been able to create, modify or abolish existing large scale structures such as stars and planets. Our Sun for example gives us a temporary window of opportunity and will eventually vanish along with the

Earth itself (we have an estimated remaining time of 5 billion years on Earth). Our technology advances may save us from such extinction one day by transporting life to another benign planet. As Earth ceases to exist, organisms by then might develop ways and means to transfer themselves to other oases that can sustain life suited to them. Otherwise, life as we know it will have no option but to vanish along with the Earth itself.

We have seen that we have stepped into an era where technology tools can now change and create organisms on Earth. So the products of Technology Bang are now in a position to alter the products of Life Bang. We have reached this milestone 4.5 billion years after the formation of Earth. Sometime in the future, we might cross the next milestone. In other words, our technology tools can start changing and creating the large scale structures in the universe. So far, we have taken them for granted as they were formed and evolved through fundamental forces of nature (natural purposeless innovation). As our technology tools are purpose oriented and deliver intended purposes, this next milestone would entail creating or changing these large scale structures in the universe toward specific purposes. At this point, we are far from accomplishing this possibility, yet it might also be a matter of time. We are still exploring the existing universe and trying to understand it. But one day, we might start shaping it rather than taking it for granted.

When one ponders the prospects for further innovation, there are myriad opportunities ahead of us. We have to a large extent mastered the skills of transforming the non-living matters (material existence on Earth) and are now quickly learning to master the living matters (organisms) as well. The Big Bang and stellar deaths have bestowed us the raw materials (chemical elements on Earth as our toolbox) which we utilize for innovation on Earth. We mix and match the parts in our toolbox to create new parts and components (composite parts) and connect them in countless ways and shape them to deliver new purposes. This has been an ongoing struggle in the form of organisms and technology tools and will continue to do so. We have also learnt to combine the forces of organisms and technology tools recently which takes us to a new level for innovation. So far, we have been limited to Earth in terms of locale.

Maybe as they say, sky is the limit! Barring unusual circumstances, we will continue our innovation journey with an incessantly abundant toolbox and in an expanded locale in the future. The imagination of the organisms will determine the pace, the outcomes and the ever changing destination in this journey. As transient passengers in this ongoing innovation journey, we witness and hopefully contribute to it.

Bibliography and Suggestions for Further Reading

Agutter, Paul S., and Wheatley, Denys S. *About Life: Concepts in Modern Biology.* Springer, 2007.

Alberts, Bruce., Bray, Dennis., Hopkin, Karen., Johnson, Alexander., Lewis, Julian., Raff, Martin., Roberts, Keith., and Walter, Peter. *Essential Cell Biology.* Garland Science, 2010.

Alon, Uri. *An Introduction to Systems Biology.* Taylor & Francis Group, 2007.

Angelo, Joseph A. Jr. *Life in the Universe.* Facts on File, 2007.

Angelo, Joseph A. Jr. *Quantifying Matter.* Facts on File, 2011.

Angelo, Joseph A. Jr. *Solid Matter.* Facts on File, 2011.

Arthur, W. Brian. *The Nature of Technology.* Allen Lane, 2009.

Asimov, Isaac. *Building Blocks of the Universe.* Abelard-Schuman Publishers, 1957.

Aslaksen, Erik W. *Designing Complex Systems.* Taylor and Francis Group, 2009.

Avise, John C. *Inside the Human Genome.* Oxford University Press, 2010.

Balbi, Amadeo. *The Music of the Big Bang.* Springer, 2007.

Barrett, Lucy W., Fletcher, Sue., and Wilton, Steve D. *Untranslated Gene Regions and other Non-coding Elements.* Springer, 2013.

Barrett, Sondra. *Secrets of Your Cells.* Sounds True, 2013.

Barrow, John D. *New Theories of Everything.* Oxford University Press, 2007.

Barrow, John D. *The Artful Universe Expanded.* Oxford University Press, 2005.

Bars, Itzhak and Terning, John. *Extra Dimensions in Space and Time.* Springer, 2010.

Battail, Gerard. *Information and Life.* Springer, 2014.

Beckerman, Martin. *Molecular and Cellular Signaling.* Springer, 2005.

Beech, Martin. *The Large Hadron Collider.* Springer, 2010.

Boyd, Richard N. *Stardust, Supernovae and the Molecules of Life.* Springer, 2012.

Brekke, Pal. *Our Explosive Sun.* Springer, 2012.

Brooker, Robert J. *Concepts of Genetics.* McGraw Hill, 2012.

Browne, John. *Seven Elements that Have Changed the World.* Weidenfeld & Nicolson, 2013.

Bryson, Bill. *A Short History of Nearly Everything.* Random House, 2003.

Bux, Faizal. *Biotechnological Applications of Microalgae.* CRC Press, 2013.

Calle, Carlos I. *The Universe: Order without Design.* Prometheus Books, 2009.

Carey, Nessa. *The Epigenetics Revolution.* Columbia University Press, 2012.

Carlson, Robert H. *Biology is Technology.* Harvard University Press, 2010.

Carr, Bernard. *Universe or Multiverse.* Cambridge University Press, 2007.

Carroll, Sean. *The Particle at the End of the Universe.* Dutton, 2012.

Carroll, Sean B. *The Making of the Fittest.* W. W. Norton & Company, 2007.

Carroll, Sean B. *Endless Forms Most Beautiful.* W. W. Norton & Company, 2005.

Carroll, Sean B., Grenier, Jennifer K., and Weatherbee, Scott D. *From DNA to Diversity.* Blackwell Publishing, 2005.

Chaisson, Eric and McMillan, Steve. *Astronomy Today.* Pearson Addison-Wesley, 2011.

Chaisson, Eric. *Cosmic Evolution.* Harvard University Press, 2001.

Chaisson, Eric. *Epic of Evolution.* Columbia University Press, 2005.

Chela-Flores, Julian. *The Science of Astrobiology.* Springer, 2011.

Church, George., and Regis, Ed. *Regenesis.* Basic Books, 2012.

Clark, David. *Molecular Biology: Understanding the Genetic Revolution.* Elsevier Academic Press, 2005.

Clark, David P., and Pazdernik, Nanette J. *Biotechnology: Applying the Genetic Revolution.* Elsevier Academic Press, 2009.

Clark, David P., and Pazdernik, Nanette J. *Biotechnology: Academic Cell Update.* Elsevier Academic Press, 2012.

Close, Frank. *The New Cosmic Onion.* CRC Press, 2007.

Cohen, William W. *A Computer Scientist's Guide to Cell Biology.* Springer, 2007.

Comins, Neil F. *Discovering the Essential Universe.* W. H. Freeman and Company, 2009.

Comins, Neil F. and Kaufmann, William J. III. *Discovering the Universe.* W. H. Freeman and Company, 2008.

Condie, Kent C. *Earth as an Evolving Planetary System.* Academic Press, 2011.

Courtillot, Vincent. *Evolutionary Catastrophes: The Science of Mass Extinction.* Cambridge University Press, 2003.

Cox, Brian. *Wonders of the Universe.* HarperCollins Publishers, 2011.

Cowan, Marjorie Kelly. *Microbiology: A Systems Approach.* McGraw Hill, 2012.

Cracraft, Joel., and Donoghue, Michael J. *Assembling the Tree of Life.* Oxford University Press, 2004.

Cranford, Jerry L. *From Dying Stars to the Birth of Life.* Nottingham University Press, 2011.

Darling, David. *Gravity's Arc.* John Wiley & Sons, 2006.

Davidson, Eric H. *The Regulatory Genome.* Academic Press, 2006.

Davies, Paul and Gregersen, Niels Henrik. *Information and the Nature of Reality.* Cambridge University Press, 2010.

Davies, Paul. *The Eerie Silence.* Allen Lane, 2010.

Decker, Heinz., and Van Holde, Kensal E. *Oxygen and the Evolution of Life.* Springer, 2011.

De Duve, Christian. *Singularities: Landmarks on the Pathways of Life.* Cambridge University Press, 2005.

Denny, Mark., and McFadzean, Alan. *Engineering Animals: How Life Works.* Harvard University Press, 2011.

Dick, Steven J. and Lupisella, Mark. *Cosmos and Culture.* U.S. Government Printing Office NASA, 2009.

Dickerson, C.E., and Mavris D. N. *Architecture and Principles of Systems Engineering.* Taylar and Francis Group, 2010.

Dieter, George E. and Schmidt, Linda C. *Engineering Design.* McGraw Hill, 2009.

Eales, Stephen. *Origins.* Springer, 2007.

Encrenaz, Therese. *Searching for Water in the Universe.* Springer Praxis, 2007.

Enger, Eldon D., Ross, Frederick C., and Bailey, David B. *Concepts in Biology.* McGraw Hill, 2012.

Fix, John D. *Astronomy Journey to the Cosmic Frontier.* McGraw-Hill, 2008.

Fox, Karen C. *The Big Bang Theory.* John Wiley & Sons, 2002.

Freeman, Ken and McNamara, George. *In Search of Dark Matter.* Springer,

2006.

Gamow, George. *A Planet Called Earth.* The Viking Press, 1963.

Garfinkle, David and Garfinkle Richard. *Three Steps to the Universe.* The University of Chicago Press, 2008.

Gargaud, Muriel., Martin, Hervé., and Claeys, Philippe. *Lectures in Astrobiology Vol I and II.* Springer, 2007.

Gargaud, M., Martin, H., López-García P., Montmerle T., and Pascal R. *Young Sun, early Earth and the Origins of Life.* Springer, 2012.

Garratt, James. *Design and Technology.* Cambridge University Press, 1993.

Garrett, Reginald H., and Grisham, Charles M. *Biochemistry.* Brooks / Cole, Cengage Learning, 2010.

Gasperini, Maurizio. *The Universe Before the Big Bang.* Springer, 2008.

Gater, Will. *The Cosmic Keyhole.* Springer, 2009.

Glendenning, Norman K. *After the Beginning.* Imperial College Press, 2004.

Glendenning, Norman K. *Our Place in the Universe.* Imperial College Press, 2007.

Glick, Bernard R., Pasternak, Jack J., and Patten, Cheryl L. *Molecular Biotechnology: Principles and Applications of Recombinant DNA.* ASM Press, 2010.

Goldberg, Dave. *The Universe in the Rearview Mirror.* Dutton, 2013.

Gomperts, Bastien D., Tatham, Peter E. R., and Kramer, Ijsbrand M. *Signal Transduction.* Academic Press, 2003.

Gregory, T. Ryan. *The Evolution of the Genome.* Elsevier Academic Press, 2005.

Gribbin, John. *Alone in the Universe.* John Wiley & Sons Inc., 2011.

Gribbin, John. *In Search of the Multiverse.* John Wiley & Sons Inc., 2009.

Gribbin, John. *The Universe.* Allen Lane, 2007.

Grier, Jennifer A. and Rivkin, Andrew S. *Inner Planets.* Greenwood Press, 2010.

Goodsell, David S. *Bionanotechnology: Lessons from Nature.* Wiley-Liss, 2004.

Halley, J. Woods. *How Likely is Extraterrestrial Life.* Springer, 2012.

Halpern, Paul and Wesson, Paul. *Brave New Universe.* Joseph Henry Press, 2006.

Halpern, Paul. *Collider.* John Wiley & Sons Inc., 2009.

Halpern, Paul. *Edge of the Universe.* John Wiley & Sons Inc., 2012.

Hancock, John T. *Cell Signaling.* Oxford University Press, 2010.

Hanslmeier, Arnold. *Water in the Universe.* Springer, 2011.

Happian-Smith, Julian. *An Introduction Modern Vehicle Design.* Reed Educational and Professional Publishing, 2002.

Hartwell, Leland H., Hood, Leroy., Goldberg, Michael L., Reynolds, Ann E., Silver, Lee M., and Veres, Ruth C. *Genetics: From Genes to Genomes.* McGraw Hill, 2008.

Hawking, Stephen. *Black Holes and Baby Universes.* Bantam Books, 1993.

Hawking, Stephen. *A Brief History of Time.* Bantam Books, 1996.

Hawking, Stephen. *A Briefer History of Time.* Bantam Books, 2005.

Hawking, Stephen. and Mlodinow, Leonard. *The Grand Design.* Bantam Books, 2010.

Hillis, David M., Sadava, David., Heller, H. Craig., and Price, Mary V. *Principles of Life.* Sinauer Associates, 2012.

Hooft, Gerard't. *Playing with Planets.* World Scientific Publishing, 2008.

Howland, Martin. *Deep-Water Coral Reefs.* Springer Praxis, 2008.

Impey, Chris. *Talking about Life: Conversations on Astrobiology.* Cambridge University Press, 2010.

Inglis, Mike. *Astrophysics is Easy.* Springer, 2007.

Irwin, Louis Neal., and Schulze-Makuch, Dirk. *Cosmic Biology.* Springer Praxis, 2011.

Irwin, Patrick. *Giant Planets of Our Solar System.* Springer, 2009.

Jayawardhana, Ray. *Strange New Worlds.* Princeton University Press, 2011.

Johnson, George., and Losos, Jonathan. *Essentials of the Living World.* McGraw Hill, 2008.

Jones, Phillip. *Science Foundations The Genetic Code.* Infobase Publishing, 2011.

Kamrani, Ali K., and Azimi, Maryam. *Systems Engineering Tools and Methods.* Taylor and Francis Group, 2011.

Kaneko, Kunihiko. *Life: An Introduction to Complex Systems Biology.* Springer, 2006.

Karp, Gerald. *Cell and Molecular Biology.* John Wiley & Sons, 2010.

Kauffman, Stuart. *At Home in the Universe.* Oxford University Press, 1996.

Kauffman, Stuart. *Reinventing the Sacred.* Basic Books, 2008.

Kazazian, Haig H. *Mobile DNA.* Pearson Education, 2011.

Keel, William C. *The Road to Galaxy Formation.* Springer Praxis, 2007.

Kepes, Francois. *Biological Networks.* World Scientific Publishing, 2007.

Kessel, Amit., and Ben-Tal, Nir. *Introduction to Proteins: Structure, Function, and Motion.* Taylor and Francis Group, 2011.

Kitchin, Chris. *Exoplanets.* Springer, 2012.

Klug W. S., Cummings M. R., Spencer, C. A., and Palladino, M. A. *Concepts of Genetics.* Pearson Education, 2012.

Kmiec, Pawel Sariel. *Unofficial LEGO Technic Builder's Guide*. No Starch Press, 2012.

Konieczny, Leszek., Roterman-Konieczna, Irena., and Spólnik, Pawel. *Systems Biology: Functional Strategies of Living Organisms*. Springer, 2014.

Koolman, Jan., and Roehm, Klaus-Heinrich. *Color Atlas of Biochemistry*. Georg Thieme Verlag, 2005.

Kosky, Philip., Wise, George., Balmer, Robert., and Keat, William. *Exploring Engineering*. Elsevier Academic Press, 2010.

Kotz, John C., Treichel, Paul M., and Townsend, John R. *Chemistry and Chemical Reactivity*. Brooks / Cole, Cengage Learning, 2010.

Krauss, Lawrence M. *A Universe from Nothing*. Free Press, 2012.

Krauss, Lawrence M. *Atom*. Warner Books, 2001.

Kurzweil, Ray. *The Singularity is Near*. Penguin Group, 2005.

Kwok, Sun. *Stardust The Cosmic Seeds of Life*. Springer, 2013.

Kundu, Tapas K. *Epigenetics: Development and Disease*. Springer, 2013.

Lagos, Claudia Del P. *The Physics of Galaxy Formation*. Springer, 2014.

Lang, Kenneth R. *Parting the Cosmic Veil*. Springer, 2006.

Lang, Kenneth R. *The Cambridge Guide to the Solar System*. Cambridge University Press, 2011.

Lasota, Jean-Pierre. *Astronomy at the Frontiers of Science*. Springer, 2011.

Lesk, Arthur M. *Introduction to Protein Architecture*. Oxford University Press, 2001.

Levin, Frank. *Calibrating the Cosmos*. Springer, 2007.

Lewin, Benjamin. *Essential Genes*. Pearson Education, 2006.

Lidsey, James E. *The Bigger Bang*. Cambridge University Press, 2000.

Lloyd, Seth. *Programming the Universe.* Alfred A. Knopf, 2006.

Loeb, Abraham. *How Did the First Stars and Galaxies Form?.* Princeton University Press, 2010.

Luminet, Jean-Pierre. *The wraparound Universe.* A K Peters Ltd., 2008.

Lurkuin, Paul F. *The Origins of Life and the Universe.* Columbia University Press, 2003.

Madigan, Michael T., Martinko, John M., Stahl, David A., and Clark, David P. *Biology of Microorganisms.* Pearson Education, 2012.

Mandenius, Carl-Fredrik., and Bjorkman, Mats. *Biomechatronic Design in Biotechnology.* John Wiley & Sons, 2011.

Mason, Kenneth A., Losos, Jonathan B., and Singer, Susan R. *Biology.* McGraw Hill, 2011.

Mazure, Alain and Le Brun, Vincent. *Matter, Dark Matter, and Anti-Matter.* Springer Praxis, 2012.

McCray, W. Patrick. *The Visioneers.* Princeton University Press, 2013.

McGee, Glenn. *Beyond Genetics.* Harper Collins, 2003.

Meyer, Stephen C. *Signature in the Cell.* Harper Collins, 2009.

Michaud, Michael A.G. *Contact with Alien Civilizations.* Copernicus Books, 2007.

Miller, James D. *Singularity Rising.* BenBella Books, 2012.

Miller, Steve. *The Chemical Cosmos.* Springer, 2012.

Moran, Laurence A., Horton, H. Robert., Scrimgeour, K. Gray., and Perry, Marc D. *Principles of Biochemistry.* Pearson Education, 2012.

Myers, Chris J. *Engineering Genetic Circuits.* Taylor and Francis Group, 2010.

Nagasaki, Masao., Saito, Ayumu., Doi, Atsushi., Matsuno, Hiroshi., and Miyano, Satoru. *Foundations of Systems Biology.* Springer, 2009.

Nath, Biman B. *The Story of Helium and the Birth of Astrophysics.* Springer, 2013.

Newton, David E. *Chemistry of Space.* Facts on File, 2007.

Newton, David E. *Chemical Elements.* Gale, Cengage Learning, 2010.

Niehoff, Debra. *The Language of Life.* Joseph Henry Press, 2005.

Ochiai, Eiichiro. *Chemicals for Life and Living.* Springer, 2011.

Oppenheimer, Clive. *Eruptions that Shook the World.* Cambridge University Press, 2011.

Panek, Richard. *The 4% Universe.* Houghton Mifflin Harcourt Publishing, 2011.

Panno, Joseph. *The Cell: Evolution of the First Organism.* Facts on File, 2005.

Pasachoff, Jay M., Flippenko, Alex. *The Cosmos Astronomy in the New Millennium.* Thomson Brooks / Cole, 2007.

Passarge, Eberhard. *Color Atlas of Genetics.* Georg Thieme Verlag, 2007.

Petroski, Henry. *The Evolution of Useful Things.* Vintage Books, 1994.

Pevzner, Pavel., and Shamir, Ron. *Bioinformatics for Biologists.* Cambridge University Press, 2011.

Phillips, Robert W. *Grappling with Gravity.* Springer, 2012.

Pierce, Benjamin A. *Genetics: A Conceptual Approach.* W. H. Freeman and Company, 2012.

Pierce, Benjamin A. *Genetics Essentials: Concepts and Connections.* W. H. Freeman and Company, 2010.

Plaxco, Kevin W., and Gross, Michael. *Astrobiology: A Brief Introduction.* The Johns Hopkins University Press, 2006.

Polizzi, Karen M., and Kontoravdi, Cleo. *Synthetic Biology.* Springer, 2013.

Popa, Radu. *Between Necessity and Probability: Searching for the Definition and*

Origin of Life. Springer, 2004.

Potter, Steven. *Designer Genes*. The Random House Publishing Group, 2010.

Pudritz, Ralph, Higgs, Paul and Stone, Jonathon. *Planetary Systems and the Origins of Life*. Cambridge University Press, 2007.

Rauchfuss, Horst. *Chemical Evolution and the Origin of Life*. Springer, 2008.

Reece, Richard J. *Analysis of Genes and Genomes*. John Wiley & Sons, 2004.

Richards, Julia E., and Hawley, R. Scott. *The Human Genome: A User's Guide*. Academic Press, 2011.

Robinson, Richard. *Biology*. MacMillan Reference, 2002.

Rossi, Cesare., Russo, Flavio., and Russo, Ferruccio. *Ancient Engineers' Inventions*. Springer, 2009.

Russell, Peter J., Hertz, Paul E., and McMillan, Beverly. *Biology: The Dynamic Science*. Brooks / Cole, Cengage Learning, 2011.

Rutherford, Adam. *Creation*. Penguin Group, 2013.

Sadava, David., Hillis, David M., Heller, H. Craig., and Berenbaum, May R. *Life: The Science of Biology*. Sinauer Associates, 2011.

Sasselov, Dimitar. *The Life of Super-Earths*. Basic Books, 2012.

Sawai, Hidefumi. *Biological Functions for Information and Communication Technologies*. Springer 2011.

Schulz, Wolfgang Arthur. *Molecular Biology of Human Cancers*. Springer, 2005.

Schulze-Makuch, Dirk., and Irwin, Louis N. *Life in the Universe*. Springer, 2008.

Seeds, Mike. *The Solar System*. Thomson Brooks / Cole, 2008.

Seeds, Mike and Backman, Dana. *Foundations of Astronomy*. Brooks / Cole,

Cengage Learning, 2011.

Seeds, Mike and Backman, Dana. *Horizons - Exploring the Universe*. Brooks / Cole, Cengage Learning, 2012.

Seeds, Mike and Backman, Dana. *Stars and Galaxies*. Brooks / Cole, Cengage Learning, 2011.

Segre, Gino. *A Matter of Degrees*. Penguin Books, 2002.

Shaw, Andrew M. *Astrochemistry From Astronomy to Astrobiology*. John Wiley & Sons Ltd., 2006.

Shrewsbury, Stephen B. *Defy Your DNA*. 10 Finger Press, 2013.

Shubin, Neil. *Your Inner Fish*. Pantheon Books, 2008.

Shubin, Neil. *The Universe Within*. Pantheon Books, 2013.

Singh, Simon. *Big Bang*. Harper Collins, 2005.

Slack, J. M. W. *Essential Developmental Biology*. Blackwell Publishing, 2006.

Smith, Moyra. *Investigating the Human Genome*. Pearson Education, 2011.

Smolin, Lee. *The Life of the Cosmos*. Oxford University Press, 1997.

Snustad, D. Peter., and Simmons, Michael J. *Principles of Genetics*. John Wiley & Sons, 2012.

Snyder, Larry., and Champness, Wendy. *Molecular Genetics of Bacteria*. ASM Press, 2007.

Sobel, Dava. *A More Perfect Heaven*. Walker & Company, 2011.

Starr, Cecie., Taggart, Ralph., Evers, Christine A., and Starr, Lisa. *Biology: The Unity and Diversity of Life*. Thomson Brooks / Cole, 2006.

Stebbing, Tony. *A Cybernetic View of Biological Growth*. Cambridge University Press, 2011.

Steinhardt, Paul J. and Turok, Neil. *Endless Universe*. Doubleday, 2007.

Stenger, Victor J. *God and the Atom.* Prometheus Books, 2013.

Stott, Carole., Dinwiddie, Robert., Hughes, David and Sparrow, Giles. *Space - From Earth to the Edge of the Universe.* DK Publishing, 2010.

Talaro, Kathleen Park., and Chess, Barry. *Foundations in Microbiology.* McGraw Hill, 2012.

Taylor, John C. *Hidden Unity in Nature's Laws.* Cambridge University Press, 2003.

Taylor, Stuart Ross. *Destiny or Chance Revisited.* Cambridge University Press, 2012.

Terzis, George and Arp, Robert. *Information and Living Systems.* The MIT Press, 2011.

Thwaites, Thomas. *The Toaster Project.* Princeton Architectural Press, 2011.

Timberlake, Karen. *Chemistry: An Introduction to General, Organic, and Biological Chemistry.* Pearson Education, 2012.

Toedt, John., Koza, Darell., and Van Cleef-Toedt, Kathleen. *Chemical Composition of Everyday Products.* Greenwood Press, 2005.

Tro, Nivaldo J. *Chemistry in Focus: A Molecular View of Our World.* Brooks / Cole, Cengage Learning, 2009.

Tsonis, Panagiotis A. *Anatomy of Gene Regulation.* Cambridge University Press, 2003.

Ulmschneider, Peter. *Intelligent Life in the Universe.* Springer, 2006.

Veltman, Martinus. *Facts and Mysteries in Elementary Particle Physics.* World Scientific Publishing, 2003.

Venter, J. Craig. *Life at the Speed of Light.* Viking Penguin Group, 2013.

Voet, Donald., and Voet, Judith G. *Biochemistry.* John Wiley & Sons, 2011.

Vogel, Carl. *Build Your Own Electric Motorcycle.* McGraw Hill, 2009.

Xie, Yubing. *The Nanobiotechnology Handbook.* Taylor and Francis Group, 2013.

Wagner, Andreas. *The Origins of Evolutionary Innovations.* Oxford University Press, 2011.

Waite, Nindl Gabi., and Waite, Lee R. *Applied Cell and Molecular Biology for Engineers.* The McGraw-Hill Companies, 2007.

Walhout, Marian A. J., Vidal, Marc., and Dekker, Job. *Handbook of Systems Biology.* Academic Press, 2013.

Ward, Peter D., and Brownlee, Donald. *Rare Earth.* Copernicus Books, 2000.

Wassenaar, Trudy M. *Bacteria: The Benign, the Bad, and the Beautiful.* John Wiley & Sons, 2012.

Weaver, Robert F. *Molecular Biology.* McGraw Hill, 2012.

Weinberg, Steve. *The First Three Minutes.* Fontana Paperback, 1976.

Wharton, David A. *Life at the Limits.* Cambridge University Press, 2002.

Wickramasinghe, Janaki., Wickramasinghe, Chandra., and Napier, William. *Comets and the Origin of Life.* World Scientific Publishing, 2010.

Wolfenstein, Lincoln and Silva, Joao P. *Exploring Fundamental Particles.* CRC Press, 2011.

Woolfson, Michael. *The Formation of the Solar System.* Imperial College Press, 2007.

Woolfson, Michael M. *On the Origin of Planets.* Imperial College Press, 2011.

Zhang, Aidong. *Protein Interaction Networks.* Cambridge University Press, 2009.

Zhao, Huimin. *Synthetic Biology: Tools and Applications.* Academic Press, 2013.

Zimmer, Carl. *A Planet of Viruses.* The University of Chicago Press, 2011.

Zubay, Geofrrey. *Origins of Life on the Earth and in the Cosmos.* Academic Press, 2000.

ABOUT THE AUTHOR

Okan Geray holds double major B. S. degrees in Industrial and Computer Engineering from Bosphorus University, an M.S. degree in Electrical Engineering and a Ph. D. degree in Systems and Control Engineering from University of Massachusetts at Amherst in US. He has published journal and conference papers in systems engineering and information systems management and taught in various Universities as an Adjunct Lecturer.

He has more than 15 years of experience in management consulting in various industries. He has consulted for a number of organizations in Netherlands, France, Italy, South Africa, Turkey and Dubai. He has worked in A.T. Kearney global management consulting firm as a member of its global Telecommunications and High Technology practice core team for 7 years before he joined Dubai eGovernment in 2002. He is currently working in Dubai Smart Government Department as a Strategic Planning Consultant.

www.ingramcontent.com/pod-product-compliance
Lightning Source LLC
Chambersburg PA
CBHW051715170526
45167CB00002B/664